WATER:
Warnings and Rewards

WATER:

Warnings and Rewards

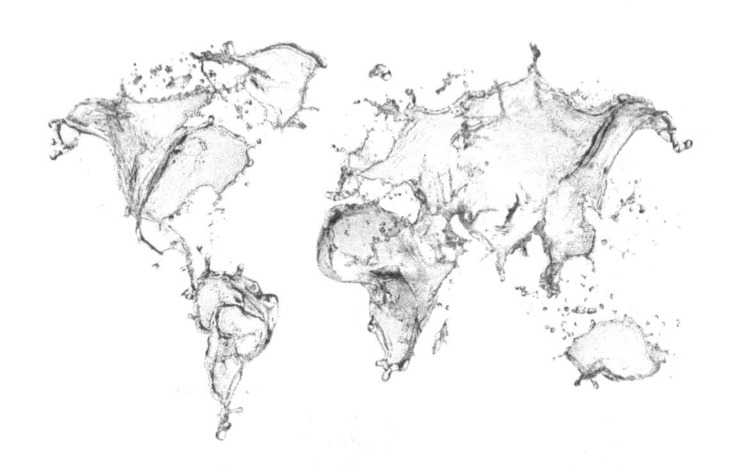

DR. THOMAS J. CALI

MILL CITY PRESS

Mill City Press, Inc.
2301 Lucien Way #415
Maitland, FL 32751
407.339.4217
www.millcitypress.net

Paperback ISBN-13: 978-1-6628-5459-0
Ebook ISBN-13: 978-1-6628-5460-6

TABLE OF CONTENTS

Water: Warnings and Rewards

— constant danger and consequent benefit, was experienced, researched, written, and referenced by Dr. Thomas J. Cali.

This book is for everyone who has ever experienced a water emergency. Whether hyper like flooding, ice and snow everywhere, glaciers and icebergs the size of some states, or conversely like severe droughts, more soil and mineral losses, changing plant nutrient and water needs, and continued variable or sometimes difficult access to problem solutions. Nutrient need, and farther or longer needs wed to further, more advanced yet increasing life denuding desertification.

Seventy percent of Earth is covered with water. The other 30 percent of Earth is covered in land. Thirty percent of the land is desert. Almost 10 percent of the Earth's land is desert or threatening to become desert. More land becomes desert as rainforests disappear. This planet is losing living space for life forms.

For coastal humans caught in the middle, water is life or death. You decide. If we don't fix the world's problems with water, there is little need to fix the myriad of other serious problems on this planet. Let's fix what can be fixed regarding our treatment of water. Start there. Make that better.

If you have ever been in a hospital or nursing home, you have seen a little water pitcher on each patient's bedside table. That water is not commonly used but is discarded from millions of sick people's rooms, minimally once a shift or once every day.

More water is used for humanity's everyday needs. That is at least a quarter million gallons of fresh drinking quality water used daily. Are you getting thirsty yet? Many other folks are. Get flowers and water them.

My passion is for you to continually get the best water from a world that is worsening water quality issues with its actions every day. Someone somewhere is doing something they shouldn't do, or not doing something they should do. Don't settle. Do what your heart tells you to do. Let your brain and mind, intelligence, intuition, and other outcomes, in short, intermediate, long, and forever time zones lead the way. One daily pitcher refill and we are over a quarter million pitchers or more than millions of gallons of water used daily. For some people, that water helps keep them alive. Follow your intuition. Make water better before you drink, use it for secondary purposes, or waste it.

When you co-exist with rival groups or tribes, they may represent a unified strong adversary, with your water as their target. There are many more of them than each individual you the defender. If water is the accidental pollution result, but forever lasting clean water need and commitment, who knows how long they will consciously or unwittingly oppose your water requirements. Water really is most everywhere. Much of it owned or restricted by regulation.

Water inside and outside are both threatened somewhere along a timeline no one knows or sees. All people have an estimate, a guess. Everyone is individually, somehow aware of time winding down. You subliminally determine your own demise and somehow from a health or bad health perspective effect your likely ending. You don't need antagonism from the environmental chemicals we cannot see or escape from. Those toxins need elimination, not just moderation.

These words are first of all, a thank you for reading. Without you there would be no need for water talk.

For everyone reading this everywhere, even those unborn when it was written, I have a special commitment to and connection with all future water, time, and you. I trust we will see each other and be here for some prospective finite time together. Just differently.

Thanks for booking. I sincerely write to help a dying world survive. You may be among the last humans to hold a book in your hands and leave a DNA trail behind with no people left to notice, follow, listen, read, or react. Are you the last contacts I will experience? Are your faces the last I will see on my journey through the revolving door, disappearing from where you and we are now? Wherever I'm going, let me say hello proactively, and goodbye conclusively, before vanishing water sources force human finality beyond people's surprise or control.

This information is more critically important to you than me. I may be already gone as you read this. Your very survival as individuals and as a segmented, world-wide, multi-societal, life-swarm depend on the presence or absence of enough uncontaminated fresh agua fria (cool water) to sustain the world's life forms and planetary landscapes.

Water : Warnings and Rewards is dedicated to everyone reading, watching, or listening. It's the learning path and refresher course for water issues throughout life's past, current, and future years. Share this book. Give it away when you're done reading. Spread the word. Water sparks healthy life. As such, water is the only completely safe and essential drug.

Water causes growth, threatening emergencies, and continually life ending disasters worldwide. It's even so in deserts, mostly from too little water and rarely because of too much. Water contamination causes more problems than natural troubles do.

This book makes you water aware, details benefits, and warns you of present and impending water risks.

The warning continues. Save yourself. Watch for beneficial and harmful water actions. Water system damage causes contaminations ranging from harmless to fatal. Water borne infectious diseases can produce rapid fatal outcomes. Each season water quality worsens, and in some areas it happens multiple times. I live in a desert getting fewer than five inches of rainwater a year. While living in the Eastern U.S., some single storms produced more water. Some years rainfall here is under one and one-half inches. There's not much water in one and one-half inches. Rainwater harvesting is a specialty unto itself. Access information from someone with that expertise before needs arise. We don't get snow. Winter holiday temperatures sometimes approach eighty- or ninety-degrees Fahrenheit.

A predicted cold spell means the high temperatures fell from eighty-five degrees today to seventy-five tomorrow. Seventy-five degrees is less warm, not cold. I see no winter coats. I haven't seen rainstorms for years. Rainfall happening here can mean drops of water on the windshield several times each year. Windshield wipers clean dust from the windows. Still no showers. Some weather forecasters talk about rain chances every day, while rain rarely happens. Often, they are sure of showers happening in the vicinity but mention the water will not touch the ground. People sometimes see a dark cloud passing with what looks like water falling but not reaching the dirt. I see no puddles anywhere. All the drainage washes are dry and have been for years. I have never seen a wash with running water, puddles, or even wet soil. The reason why is obvious. No rain.

When it does rain, there is always associated auto wreckage and loss of human life.

The local weather service daily predicts chances for rainfall, often less than a five percent risk. They forecast rain for days, weeks, months, and sometimes seasons or years ahead. Often none of those future predictions produce rain. The chances are under fifty percent and often below twenty or ten percent. A ten percent chance for rain is a ninety percent chance for no rain. That's never the headline or starting story on the news. When it rains, it is always the leading news story. They always predict occasional hope. Hope leaves life thirsty and the area dry. When rain happens it's a brief shower high in the mountains. Most mountains are so jagged and severe no one lives there, not even close. There are no roads. There are very few places flat enough to land a plane or even a helicopter. Humidity is often measured in single percentage digits for weeks and months, or possibly years. When humidity is that low you feel severe increasing thirst, signaling dehydration. Exposure to low humidity breezes makes you dryer. Dry wind makes you crave a drink of water. Every living thing is always thirsty. Increased thirst feelings are a more dependable predictor of no humidity than low numbers. Weather services often report whether rain will or won't hit the ground. Won't outnumbers will. The dew point today was in single digit numbers. What does that mean? Sometimes you can see disappearing rain evaporating from the sky without ever feeling a drop. Virga.

Ever changing news information exists for everyone. Weather prediction is a service for most of this world's people. There will be more generic data for sooner, and greater imitation and real long-term information for later, on and on, to whatever wet or dry fate awaits. Humans need lots of water every day, or we will limit the number of future days life allots us. When you drink water, eat anything made, caught from, grown, cooked, served, swallowed with water, or other pseudo-water liquids, whether flat, cold, fizzy, or frozen, changing water knowledge becomes your necessary daily objective. Wherever you stop or stay, water balance remains your necessary life goal. Water information is important to make life as long or short as you hope it to be.

This manuscript is dedicated to Rachel Carson. Her work on unfinished primary and far-reaching secondary as well as tertiary objectives, compelling reflections about toxins and consuming interest added to intense worries about Earth's water and environment live on. The societal dis-focus on "forever" water is not real, aggressive, or long lasting. The rising world population devours more water in order to live. Our broken water focus puts us farther behind humankind's increasing water needs. Healthy drinking water disappears rapidly. Improvements are not perfect, just better than before. That's why there will always be new software.

This book is also devoted to everyone dead or dying, or expecting to live on, despite water emergencies. Whenever or wherever you are reading this, today's global mortality from water issues is greater than yesterday's. Improvements certainly occur, but they are far outnumbered by the at risk or seriously affected population.

PROLOGUE

Rachel Carson wrote *Silent Spring* (Carson, Rachel, *Silent Spring*, Mariner Books, New York, 2002). The first edition was published in 1963. *Silent Spring* warned people about thousands of present unknown harmful chemicals and numerous known toxins from everywhere in all forms and locations present in the water. Some chemical and/or biological contaminants she discussed are still problems in today's drinking water, and the places we live, both in Earth's environment, and for most life forms. Thousands of new toxic synthetic materials are now used. Very few have been effectively tested for safety, in my opinion. Most substances, new and old, haven't been thoroughly evaluated.

I prefer to see more long and short-term human research observations and evaluations of both toxicity in various challenges and better efficacy considerations. Hope lives on wrestling with reality.

Each year over fifteen hundred new chemicals, about six daily, are marketed and added to the largely inexact, imprecisely tested, and far from proven safe, long-term chemical mix the United States uses. The chemicals get into the water. An unknown number of people die daily from contaminants in water, food, air, and other toxic environmental substances. Causation is difficult to objectively prove except in acute situations. Serious common microorganisms like those causing MRSA, SARS, and MERS infections have caused local or limited problems, and extensive often life ending infections. Corona virus has caused infections in most countries throughout the world. Each type of infection has caused and still causes human deaths everywhere.

Acute situations often end before causes are discovered, identified, evaluated, or fixed. Death may have already happened, perhaps before the cause and cure was identified and used where you are and elsewhere around the world.

Every person, human residence, and healthcare setting uses water. Medicines and other things including hydration solutions are given in or as distilled water administered intravenously, intramuscularly, subcutaneously, or otherwise. These products are sterile without any agents that can cause infection. Hospital acquired infections are sometimes caused by the lack of proper sterile technique by healthcare workers preparing or administering the solution. Except for distilled water, anything you drink may be contaminated. Water distillation is one barrier between clean health, toxic illness, and the sickness or water-borne illness universe between. Only you can save you. Everything you read here is critically important. All the information should be <u>underlined</u>, CAPITALIZED, in "quotes," or in **bold type**. Emphasis can't be too omnipresent or important.

More information about water has become available since the publication of Water: Warnings and Rewards. How much of what you believe about water, whether good or bad, is proven by ongoing, independent, prospective, confirmed, objective, and repeated research? Is current water from anywhere proven clean and safe?

It would be excellent if the water you drink, use to raise animals, and cook with made you healthy. Contaminated water can make you sick. It would be best if streams, rivers, ponds, lakes, and oceans were clean, and if beaches never had toxic or pollution reasons to close and you could swim anywhere. Everyone has experienced one or more of these water bodies being contaminated.

It would be great if the water you shower in and wash with is safe. It would be wonderful if the water falling onto you from the sky was non-toxic. Most rain contains mercury, lead, aluminum, perchlorate, and other harmful pollutants it obtains falling through the sky or from surface water.

Read this book and you will find something in water possibly making you sick. Contaminated water may be killing you. Someone you know may have already been harmed by water. There is no information in this book predicting what will happen tomorrow. No one can tell you what tomorrow's water will bring, or what the weather will be. No one can tell you whether there will be a tomorrow.

What you read here is happening now. It's hard enough to figure out today.

The contamination in water and water borne diseases causes harm to people. Don't let people keep doing what they have always done to water. Don't let people continue to sacrifice the only planet we have by poisoning its water.

INTRODUCTION

Recycled water! What wait, isn't all water recycled from someone, something, or somewhere? Every drop you have ever touched, tasted, or seen up close as it blew into your eyes or flew onto your face during a storm, others have experienced every way you can imagine, and some ways you have not dreamed up yet. My mind is constantly aware of every raindrop, stream, creek, river, bay, lake, ocean, frozen island, glacier, icecap, iceberg, puddle, happy and scary concept about the world's water and what is further happening to it. Hopefully, there is some completely clean water left somewhere. May your children and grandchildren learn to swim outside in surface water without becoming sickened or poisoned.

When children learn to swim in a pool, can they safely look forward without fear to effects from added chlorine, combination chlorinated compounds, and other synthetic chemicals, whether safe, dangerous, deadly, or producing largely unknown outcomes good or bad?

Probably both.

Who knows whether clean water anywhere will survive political changes or toxin contamination realities from everywhere. Toxins, like perchlorate and others from war in Ukraine, will eventually encircle the globe. First in the air we all breathe, then in the water we drink. Let's hope eventually is a very long time from your now.

My first memory of a serious water warning came from relatives and was supported by neighbors in my hometown.

Main Street went from the Eastern town's edge, less than one mile straight West to the opposite border, the Delaware River. Children were often discovered in the water where they shouldn't have been.

During my childhood, the town lost only a few children from drowning accidents or diseases like cancer. Every loss was equally important and cannot be over-rated or over-stated. The Delaware River extended a couple miles across, from New Jersey to Delaware. The river was more than one hundred feet deep and two miles wide in places. Pollution prevented anyone from swimming. It still does. Three separate piles of large boulders were placed in deep water twenty or thirty feet from the New Jersey shoreline, where the current moving out toward deep water is strongest. In the small town's early years, three children died there before rocks were placed in the water, as subliminal warnings to all.

Each time kids drowned, the town stopped breathing and started screaming. Everybody felt it. Everyone in town knew someone who knew at least one of the three original kids. Some people knew all three.

More children would die here before me.

The town never recovered from the tragedies. Those three separate piles of stones in the water close to the shore serve as the town's memory. Children were not allowed on the beach or in the water near the stones. The town's adults enforced the ban on anyone's kids seen playing nearby. The rocks mark the disasters. Both stones and people are constant reminders. Look there and see the stones even today. All the town's kids were told the tale as a warning. Some people, mostly children, still hear past screams in current dreams and nightmares. We all heard the stories.

The water is in worse condition now. If you ever tried to do anything using water and it did not work out, try it again with purer water. Better yet, try it again using pure distilled water and see what good happens, and what bad doesn't happen.

Rather than struggling with chronic disease prevention, avoid acute disease causation and progression from contaminated water.

Bad water is an unfixable, worsening, pre-existing condition affecting everyone. It causes individual infections, worsens infections into epidemics, and produces pandemics affecting the Earth. Rainwater acquires the heavy metal toxins mercury, lead, toxic perchlorate, and others as it falls to the ground. Clean rainwater times out immediately when it hits the ground and mixes with poisons people leave on the ground or in the water. Surface water makes all these poisons more available and produces negative effects.

Big numbers are their own numeric significance. One thousand milligrams sounds like it deserves more attention than one gram, but it doesn't.

Many people are killed by current drinking water systems throughout the world. More people are killed whenever anything else is put into veins or other blood vessels without proper sterilization, infection prevention processes, and other needed personal protection pretreatments. Since pure water gives, and bad water takes life, water is already challenging enough to mandate continual mindful attention. Humans need to stop using bad water and/or making good water worse before use. Use only good water while continuing to make bad water better. Many species are currently becoming extinct as dehydration and desertification claim more parts of the Earth.

This is not a novel, not a "how to" book. This book tells what will happen if people keep doing the wrong things and avoid doing the right things for water, planet, and people. Follow a good combination of innate "smarts" and time-tested intuition. Live your life. Ease the stress from how and where you live by improving water at all or any levels.

This book will be used by everyone in the world. These words honor any action people take to help water anywhere. It's already too late for many people somewhere.

I grew up in Southern New Jersey in the 1950's and 60's. To say it was a simpler time, understates my early childhood in small town America. I grew up when kids played outside all day, every day, all summer, all the time. Kids grew up without the Internet, video cameras, video games, or cell phones. Some phones had party lines that were no party. The outdoors, including streams, lakes, oceans, beaches, fields, and woods were our playground. I grew up on the Jersey Shore and loved it. Many things have changed. Cancer rates in New Jersey have increased since my childhood, as reported by the New Jersey Department of Health in 2013.[1] Salem County, where I grew up, has the highest mortality rates in the state, according to a 2014 report from the Centers for Disease Control and Prevention.[2]

I have always had an interest, bordering on obsession, in health and people. This led me to a career trying to help people maintain, and where possible, restore their health. I attended what was then called Philadelphia College of Pharmacy and Science (PCP&S), founded in 1821. It is now known as University of the Sciences. When I graduated in 1972, it was a time when Pharmacy was poised to make a significant transformation. A doctoral degree in Clinical Pharmacy was introduced at leading Colleges of Pharmacy across the nation, beginning in California. Clinical pharmacists differ from bachelor's degree level pharmacists in that a clinical pharmacist starts their training with a bachelor's degree in pharmaceutical science, biology, or a related subject. They continue their didactic and clinical training, usually including an original research project, often published, until they have earned their doctorate degree in clinical pharmacy. Clinical pharmacists have been trained with in-depth knowledge of medications and often consult with physicians and other health care professionals on drug therapy and usage. Those interested in becoming clinical pharmacists can specialize in therapeutic areas, as do physicians.

I was quick to become part of this exciting, expanded role in Pharmacy. I received my doctorate in Clinical Pharmacy from Philadelphia College of Pharmacy and Science. After completing my degree, I held a joint faculty appointment at PCP&S and the University of Pennsylvania as an assistant professor, as well as holding a position within the hospital's Pharmacy Department. During this time, I received post-doctoral specialty training in Infectious Diseases, Nephrology, Psychiatry, and Geriatrics. I became an assistant professor at the University of Maryland School of Pharmacy in the 1980's, where I was named "Professor of the Year." After my academic career, I pursued a career as a consultant pharmacist in long-term care. I worked in the greater Philadelphia area for a private consulting firm, then moved to a larger, geriatric firm in the Tidewater area of Virginia.

In my consulting role, I became active in the professional society for consultant pharmacists–the American Society of Consultant Pharmacists (ASCP). I was recognized by this national, professional organization in 1994, by being awarded the prestigious George F. Archambault Award, the Society's highest honor. It is conferred annually to one individual for their outstanding contributions to consultant and senior care pharmacy. I have earned fellow status with ASCP, allowing me to include FASCP in my credentials.

I went into private practice in the early 2000's. Being in private practice has allowed me the flexibility to choose the cases I accept, and to work directly with sick people, a/k/a patients. I have chosen to be in private practice to meet the needs of people within my skill sets. I have consulted in North America, China, Africa, and Europe. My practice allows me to do things needing to be done, rather than participate in a reimbursement driven employer healthcare system. I remain active in private practice.

Through many years of clinical training and decades of person-care practice and experience, I saw the effects our planet's contaminated water could wreak on my patients and friends, some of whom became patients. These involvements, both good and bad, led me to advance, cultivate, and grow my interest and expertise in the beneficial health effects of clean water and the far-reaching negative disease and sickness effects from the contaminated water supply.

Abundant information is available concerning water. Some is true, some is questionable, and some is clearly wrong. A cellular telephone network computer connection makes much of the world's known information–valid, fake, and everything in between, accessible to any user.

Good research is difficult to do. Bad research is done all the time. Even useful treatments have unwanted reactions. Just because harmful effects aren't here now, doesn't mean they weren't here before. Maybe toxic effects unseen initially will be revealed later or again though unseen before. I believe many people are being harmed by the water they drink.

Parts of six states and numerous big cities like New York, Philadelphia, and others are submerged after recent storms. The New York City subway is underwater. There was a tornado in New Jersey, my home state. There is drought continuing in the West, and a significant area in California is on fire. Perhaps most significant and frightening is that climate change deniers are starting to admit global warming and climate change might actually be a real weather "thing."

CHAPTER 1

Bad water is making humanity and the world we live in die. Distillation can save the wanting clean water world. Distilled water is the failing world's only salvation. It is late but not yet too late.

Water: Warnings and Rewards was written using decades of patient care experience with people having far ranging hydration issues, and information learned from many different physicians, scientists, professors, nurses, pharmacists, children, sick and healthy people, social workers, friends, intellectual enemies, students, and others. A sincere and heartfelt thank you to everyone. You know who you are. Always remember this work has taken years to produce and will never be complete. Water issues will never be finished. I decided to write this book because toxins are in the water now that weren't there when I was growing up, as confirmed in an Environmental Protection Agency (EPA) report published in August of 2017.[3] More importantly, I wrote this book to provide more information on how to protect yourself from contaminated water and how to best use whatever water is available to you.

Water is a subject very important to me on a personal level as well. I live on this planet with you all and your water. I live on this planet with people I love. I have seen friends and others die from water. I have a partner and children who I am seeing grow, thrive, and age, which beats all the alternatives. I expect to become older despite them attempting to kill me with laughter and psychological protests every day.

While sitting in a park outside of Philadelphia, Pennsylvania reviewing this manuscript, a Philadelphia County weed abatement employee arrived. He began to spray an amber colored herbicide on the grounds of the park along the Schuylkill River. Although upwind from the spraying, I noticed the smell and asked, “What chemicals are in the spray you are using?”

He replied, “I don’t know what this weed killer is called, but it’s safe and doesn’t smell, at least not too bad to me.”

If he didn’t know the herbicide’s name, how does he know the unknown liquid was safe? He wore no gloves, mask, or protection of any sort. I don’t believe there are any plant killer exposures safe for people, other life forms, the land, the food, the environment, and especially the water.

He was spraying green vegetation at the Schuylkill River’s edge near two people fishing. The spray hit the water covering a wide circle. He said nothing to the people fishing as he sprayed their sneakers, along with the vegetation around them. The spraying of any “-cidal” or killing compound by itself was enough reason for me to get away from the park. While some effects of this chemical are known, the range of possible adverse effects on the water, the people drinking or using it, and the plants and creatures living there are largely unknown.

Don't touch the water or drink the "Schuylkill River Punch" as locals call it. Fish caught there aren't safe to eat.[4] In my opinion, there should have been a sign warning people the fish are unsafe. This scenario made me wonder–is water any place safe? It also made me question the affects every day common practices have on the potential to do harm to the environment, our water table, and ultimately to people and other life forms.

The man was spraying the herbicide RoundUp®. All the empty containers in the truck were labeled as RoundUp®. Glyphosate is the most commonly used herbicide in America[5]. RoundUp , or glyphosate, is recognized as a probable human cancer-causing agent by the United States EPA.[5] It seems reasonable to me to conclude that if glyphosate is a probable cancer-causing agent, it can't be safe to spray into water used for drinking by people. The water contains green things, plants, and other life forms like fish, turtles, and more. All may be affected by overspray of glyphosate. I don't think the presence of a cancer-causing agent in the water is safe for them either.

Glyphosate is the chemical name for a RoundUp® herbicide component.[6] Glyphosate is one of the most commonly used herbicides in commercial farming to control weeds and clear the fields prior to planting the next crop cycle.

RoundUp® works by attacking plants in 3 different ways. It kills plants and weeds by blocking the plant's ability to absorb an essential element, manganese; by binding or chelating essential minerals the plant needs to thrive; and by allowing bacteria harmful to the plant to kill the plant in its weakened state.[7] Glyphosate also enters the soil when it is used to control weeds. That is how this chemical works.[8] Although glyphosate is intended to kill the unwanted plants, it may have the unintended effect of removing minerals from the soil of plants that are grown for human and animal consumption. As a result, the fruit, vegetables, grains, and crops that we eat, and the animals grown for food eat, may have less mineral content.[9] However, these findings are controversial and sometimes conflicting. If you believe as I do, that plants grown for human and livestock consumption may be subject to less mineralization due to commercial herbicide use, then you may conclude glyphosate decreases dietary mineral availability to people too. There are essential minerals needed by plants, animals, and humans that glyphosate removes from soil. This problem may be further compounded since those minerals aren't replaced by drinking water.

In addition to the chelation of soil and plant minerals, glyphosate also has affects on microbes living in soil. It acts on harmful and beneficial soil microbes.[10] It may interest you to know that Monsanto initially patented glyphosate as an antibiotic.[11] Glyphosate promotes the growth of harmful bacteria in order to kill weeds and kills the helpful bacteria normally present in soil to protect crops from the harmful bacteria.[12] As a result of the antibiotic effects of glyphosate, resistant strains of bacteria infecting people and animals have occurred.

Glyphosate contamination is widespread.[13] In a 2002 study by the US Geological Service (USGS), it was found that 154 water samples from 9 Midwestern States collected from 51 streams, were contaminated. Thirty-six percent of the samples detected glyphosate, and its degradation product (aminomethylphosphonic acid, AMPA) was found in 69 percent (of samples collected. Levels detected were below the maximum contaminant level (MCL) set by the EPA for glyphosate, and levels for AMPA do not have an MCL established.[14] My question is – do we want any amount of glyphosate or it's metabolite contaminants in the food we eat and the water we drink?

In my mind, it's not hard to make the connection that if a contaminant is in the soil, it could easily enter the water supply through weather extremes and storm runoff. However, glyphosate also enters and is "recycled" in water through rainfall. In a study published in the journal Environmental Toxicology and Chemistry by the USGS in 2014, it was reported that in 2007 glyphosate and its degradation product, AMPA, were detected in greater than 75 percent of air and rain samples.[15]

Wait again, what? Rain contains a potentially toxic plant killer? Rain goes everywhere, falls on everyone and everything, and is used by plants worldwide. Glyphosate contaminated rainwater may kill plants even in areas where it was not applied. To my way of thinking, that is not good.

In a 2012 publication, glyphosate, or RoundUp®, has been found in urine from people living in some German cities at levels five to twenty times higher than RoundUp® amounts allowed in German drinking water.[16] Do you know what you're drinking? In my opinion, glyphosate exposure seems to be unavoidable. In the water you drink and food you eat, you're likely to be exposed to glyphosate through soil, water, and/or rain contamination. RoundUp®, or glyphosate, contaminates water sources throughout the world.[17] That is not hard for me to believe since billions of kilograms of it is used worldwide on an annual basis.[18]

With the potential for glyphosate to contaminate the water system through many points of entry, there is much debate over the possible consequences it may have on those who come in contact with it and/or consume it in their water or food sources. The internet is full of information, some reliable, some not reliable, on this topic. Read and decide for yourself and your family.

You will read glyphosate is involved in many toxic disorders. It has been reported to be associated with general toxicities, hepato- or liver toxicity, nephro- or kidney toxicity, certain cancers, teratogenicity or birth defects, and many others.[19] Both the World Health Organization (WHO) and EPA have released statements on glyphosate and its potential link as a cancer-causing agent. The World Health Organization (WHO) has a specialized agency that investigates cancers called the International Agency for Research on Cancer (IARC). In March of 2015, the IARC reported glyphosate is classified as probably carcinogenic to humans (Group 2A).[20] In December of 2017, the EPA released a draft risk assessment for glyphosate and concluded that glyphosate is not likely to be carcinogenic to humans.[21] The EPA's assessment found no other meaningful risks to human health when the product is used according to the pesticide label. The findings of the EPA are consistent with the conclusions of scientific reviews by a number of other countries, as well as the 2017 National Institute of Health Agricultural Health Survey.[22]

How do you resolve conflicting reports from what people deem sincere and valid sources? Most of us, myself included, would like to trust regulatory agencies and organizations focused on public and/or global health. In cases like this with conflicting conclusions from credible and believable sources, I believe in order to stay ahead of the possible toxicity curve, you must limit your exposure to likely toxins.

You don't want to have one of those toxicities or expose yourself to possible cancer-causing agents. In my opinion, the burden of proof for glyphosate and safety or harm is on the person being harmed, not the manufacturer. This isn't correct. No level or harm to a person by a chemical is acceptable to me. Ask yourself what is an acceptable level of harm or risk to you, your children, and your family? Is the danger of RoundUp® exposure finally being recognized by the public in the form of a class action lawsuit filed in the United States Federal court system?[23]

As previously mentioned, RoundUp® kills everything green and not "RoundUp® ready." "RoundUp® ready" means that a crop seed has its DNA genetically modified so the plant can withstand the effects of glyphosate, one of the main herbicides in RoundUp®.[24,25,26] The plant has been genetically altered, also known as a genetically modified organism (GMO). GMO may be a more familiar or often used term.

The world's livestock populations are the largest consumers of GMO crops.[27] It is estimated that over 70 to 90 percent of GMO harvested crops are fed to food producing animals. GMO crops are likely to become even more important to animal agriculture as the global livestock population increases in response to growing demand for animal protein products.[28] More than 95 percent of food producing animals in the United States consume feed using GMO ingredients.[29] Worldwide, approximately eighty billion animals are raised and killed for food each year.

There is much controversy surrounding the use of, and consumption by, livestock and humans of GMOs.[30] It is beyond the scope of this book to address the controversy of whether or not to consume these products. My focus remains on the affects of these substances on the worldwide water system.

Organic fields near genetically modified RoundUp® ready crops may not be organic. The same RoundUp®

contaminated water flows to those crops. The same RoundUp® contaminated wind blows there. How much contamination there is depends on the weather, heat or cold, rainfall, and wind. There has been at least one court case on this matter. A recent case pitting a Canadian farmer against Monsanto was decided by the Canadian Supreme Court, demonstrating GMO seeds could be found in the fields of a farm that did not use GMO products.[31]

I don't want those substances in the foods I, or anyone eats, until genetic modification is scientifically and repeatedly proven safe for life forms. In my opinion, GMO modification hasn't been studied correctly or long enough in human or animal populations to uncover all possible toxicities, or potential genetic implications or surprises. Protracted studies, longer than ten years, have not been evaluated thoroughly in people with a variety of skills and/or physical or emotional challenges. A ten-year study of people is less than fifteen percent of our approximate predicted seventy-year life span for both females and males. Could genetic modification of life forms cause our extinction? Genetic modification lasts a long time and may not be able to be reversed in some species. Extinction lasts forever. It is logical to me that GMOs get into your body from the water you drink and the foods you eat. Let's find out and prove what good and bad things GMOs do before they're used on people with or without their knowledge.

In my opinion, genetically modified organisms are a giant food experiment, with people and animals raised for food as the test subjects. It is alarming to me that most of the human test subjects don't know they are part of an experiment.

CHAPTER 2

RoundUp® kills every green plant not "RoundUp® ready." Agent Orange killed every green plant, including trees, when used in Southeast Asia. The two products are related in that they share similar toxic effects to plants and are made by the same company, Monsanto, now owned by Bayer. Both products touched my life. Whether it's sitting in a park reviewing this manuscript or reminiscing about an old friend.

Woody will always be my friend. During the Vietnam War, Woody encountered the powerful herbicide, Agent Orange. Woody was drafted; he didn't volunteer to die. Woody survived bullets, missiles, bombs, and grenades. He lived through attacks by armed strangers. Woody died after returning home. He was in his thirties when he died from a rare muscle cancer called myosarcoma. In my view, it was caused by chemicals made in the United States and used during the Vietnam War. Woody wasn't the first person to tell me his Agent Orange story, nor will he be the last. Numerous people were harmed here, in other ally countries, and in war zones. In my opinion, the last Agent Orange story has yet to be told.

Woody's exposure was common. Agent Orange was supplied in 55-gallon drums. When empty, the containers were cut in half and holes drilled in them. Half-drums were filled with water then used as showers. Showering down on the users were leftover amounts of Agent Orange and its contaminant, dioxin. By doing their jobs serving this country, people were exposed to carcinogenic chemical toxins.

Agent Orange is an herbicide which contains two active ingredients in equal amounts–2,4-dichlorophenoxyacetic acid (2,4-D) and 2,4,5-trichlorophenoxyacetic acid (2,4,5-T), which contained traces of 2,3,7,8-tetrachlorodibenzo-p-dioxin (TCDD).[32] The dioxin TCDD was an unwanted byproduct of herbicide production.[33] Dioxins pollute the environment and are released by burning waste, diesel exhaust, some chemical manufacturing, and other processes. TCDD is the most toxic of the dioxins and is classified as a human carcinogen by the Environmental Protection Agency.[34]

The Veterans Administration (VA) has recognized certain cancers and other diseases associated with exposure to Agent Orange and other herbicides used during the war.[35] Woody's rare type of cancer is one of the cancers recognized by the VA.[36] To me, every cancer is a tragedy.

The VA also recognizes certain birth defects associated with exposure to Agent Orange by the child's parents. People not directly exposed can get birth defects from Agent Orange toxicity from parents who are Vietnam war veterans.[37] Many children in Southeast Asia were born with birth defects.[38]

Agent Orange, as well as other colorful liquids called "rainbow herbicides" were used to eliminate forest cover and crops used by the Viet Cong and North Vietnamese troops. Agent Orange was the most commonly used. From 1961 to 1971, the United States military sprayed more than 20 million gallons of herbicides over 4.5 million acres of Vietnam, Laos, and Cambodia. Unfortunately for those living in the area, crops and water sources used by the native population of South Vietnam were also affected.[39] People continue to live and vacation in Vietnam. Are they still at risk for exposure to this long-lasting toxin? I believe the answer is yes.

Dioxin in the form of TCDD persists in the environment for many years and can enter the food chain through soil, lake, and river sediments, and in the fatty tissues of fish, birds, and other animals.[40] Studies performed on laboratory animals have demonstrated that very small quantities of dioxin are highly toxic. Dioxin is universally known to be a carcinogen.[41]

Dioxins are classified by the EPA as persistent organic pollutants (POPs), which means they take a long time to break down once they are in the environment. Dioxins are known to cause cancer, reproductive and developmental problems, damage to the immune system, and can interfere with hormones.[42] Because Agent Orange is known to contain dioxin and persist in the environment for many years, contamination of water due to its use many years ago remains a concern.

It's been said "the solution to pollution is dilution."[43] In a functional sense, the Earth has only one ocean since every ocean connects with at least two others.[44] As big as the oceans are, this planet still has its limits. The Earth has a finite number of places for dilution to occur. In my opinion, poisons don't follow lines on a map. After enough time, water contaminates anywhere, and affects the ocean everywhere. Nothing put into water has boundaries. Twenty million gallons of contaminant is a big drop even for a bucket as big as the ocean. In my opinion, dilution of toxins into the water was never the answer to dispose of them.

That single ocean is part of the ecosystem caring for the human species and all life forms. Everyone and every living thing share the ocean and the ecosystem. We all have responsibility for cleaning it, and not dirtying it. We must keep water clean if the human species is to survive. We must continually protect the ecosystem that is taking care of us.

Glyphosate is not the only herbicide having the potential to contaminate our water and adversely affect our health. There are many other toxic agents used to control weeds and pests. Those chemicals enter our environment and we eat and drink them mostly without knowing. I am reminded of a scene from the classic movie *Psycho,* when a hardware clerk makes a remark about pesticides saying that pesticides are guaranteed to kill every insect in the world. The term "pesticide" covers a wide range of chemicals including, but not limited to, insecticides, fungicides, herbicides, and other agents.[45] The widespread use of various herbicides and pesticides is what scares me more than the famous shower scene in the movie *Psycho*. I believe we should be doing more to protect our endangered water system from these possible toxins.

Recent estimates report there are more than 200 million insects for every person on Earth.[46] It has always been that way and will always be so. Insects are the largest biomass on the planet.[47] To address weed and pest control we use chemicals to try and manage one issue while possibly creating another problem and harming humans in the process.

More than one billion pounds of pesticides are used in the United States each year. More than five billion pounds of pesticides are used in the world every year. That is more than a quarter pound of pesticides used for each person on Earth every year. Pesticides kill life forms aimlessly. Ideally, a pesticide should only kill the targeted pest, but not non-targeted species. However, this is not the case. Pesticides kill both helpful and harmful insects.[48] Adding to the problem with these agents, is the abuse of pesticides by using more than is either needed or directed on the label.[49] Since the focus of this book is on water contaminants, like glyphosate, widespread use and overuse of this broad range of chemicals called pesticides and herbicides has the potential to contaminate groundwater. Contamination of groundwater is of significant concern because it serves as the source of drinking water for about 50 percent of the American population. In agricultural areas, where there is higher use of pesticides, about 95 percent of people living there get their drinking water from groundwater sources.[50]

Studies have shown that water-bearing aquifers below ground can become contaminated through pesticide application to crop fields, seepage from contaminated surface water, improper disposal of pesticides and other chemicals, accidental leaks and spills, and through injection of waste material into wells. As mentioned earlier in this chapter, the EPA has set a maximum contaminant limit (MCL) for some, but not all, pesticides in drinking water.[51] Is any level proven safe? Is any level of contamination acceptable? In my opinion, there are more questions asked than answers given to these life altering queries. Another area where I am concerned is it is unknown what is considered safe by regulatory agencies when there is more than one contaminant in the drinking water. Do combinations of toxic contaminants cause more problems and/or kill more effectively? No one has sufficiently answered that question to my satisfaction.

You may be aware of a newer class of pesticides called neonicotinoids. These are popular pesticides widely used in the past 20 years to control insects like aphids and root-feeding grubs. The neonicotinoids include acetamiprid, clothianidin, imidacloprid, thiacloprid, and thiamethoxam. They act by being taken up by the plant and transported to the leaves, flowers, roots, stems, pollen, and nectar of the plant treated. The neonicotinoids remain active in the plant for several weeks, thereby protecting the crop for the entire growing season.[52] Use of these chemicals is of concern for many reasons, but two which I am most concerned about are: 1- the devastating effects they are having on the honeybee and other pollinators' populations worldwide; and 2- the potential for contamination of our water supply. Other things besides water (e.g., honey) can be contaminated by the neonicotinoids, but that is beyond the scope of this book, as is the affect on the pollinators' populations. I encourage you to read more on your own on these subjects. The EU began banning the use of these pesticides in 2017.[53] The first United States ban of the use of these agents began in Maryland in January 2018.[54]

In terms of the potential for the neonicotinoids to affect water, the Washington Post reported in April 2017, the first evidence these pesticides were found in treated drinking water.[55] A team of chemists and engineers at the USGS and University of Iowa found neonicotinoids in treated drinking water. Three neonicotinoids were present in samples of finished drinking water taken as periodic tap grabs over a 7-week period (May-July). The neonicotinoids clothianidin, imidacloprid, and thiamethoxam were detected in concentrations ranging from 0.24 to 57.3 ng/L.[56]

I believe the risk of exposure (dose) goes up as pesticide use and popularity increases around the world. Many pesticides are sprayed at schools, playgrounds, and parks, contaminating areas everyone, especially children, use. I think children may be especially susceptible to the dangers of these toxins. In my view, people who spray pesticides are the pests. I would rather live with a few bugs than be exposed to a pesticide. Malaria and other fatal diseases transmitted by insects are not common in the United States, and many can be treated with currently available medicines.

I have seen many interesting patient cases throughout my career and asked to consult on confusing complex patient care situations involving water. The following case highlights the sensitivity of certain individuals to toxins many of us may contact on a daily, weekly, monthly, or annual basis. This exposure may or may not affect us. Occasionally, this individual puzzling situation arises without easily apparent explanation. Janey is one such individual.

Long ago, I had a patient named Janey. At age four, she had a constant low-grade fever, total body skin rash causing intense itching, bone and joint inflammation causing constant severe pain, and asthma. Janey had been sick for two years, half of her young life. Her family consulted many healthcare providers, both traditional and non-traditional. None were able to provide a substantial or prolonged benefit.

Once when Janey was hospitalized for an infection, I met Janey's mother and father. I became part of the team taking care of her.

I observed Janey's activities at home, and the substances she contacted or consumed. I observed her diet. After several months, I changed Janey's drinking water from local tap water to distilled water. Several weeks later I suggested a restricted diet. She was started on a vegetarian diet of organic fruits and vegetables. After several months, her symptoms disappeared. Her body temperature, skin, bones, joints, and lung functions were normal. Janey's pain was gone.

Other foods were re-introduced into her diet weekly, slowly, and one at a time. While re-introducing food items slowly, Janey developed symptoms after eating bananas. Banana testing revealed no toxins or contaminants. Eliminating bananas from Janey's diet reversed her sicknesses. Allowing Janey to consume bananas re-produced the same diseases. Testing the banana peels revealed a half dozen toxic chemicals.

My best guess is, while peeling bananas, Janey absorbed toxic poisons through her skin into her bloodstream making her sick. Washing and peeling Janey's bananas before she ate them, and keeping her drinking distilled water, avoided all previous diseases, and kept her healthy.

I can't explain why she was more susceptible to the chemical adverse events than either me, her parents, siblings, or any other kids in her school experienced with identical food and toxic exposures. She may have been more susceptible than most of you reading this book. I can only share what my experience was with this one patient. It's experiences like this that make me leery of consuming non-organic fruits and vegetables. It also teaches me that when I see one patient get better and recover completely, I have only seen a one-person recovery and may never see it again. One case response cannot be generalized to a series of patients or even one more identical case.

Although, beyond the scope of this book I encourage you to become aware of the "dirty dozen" list of fruits and vegetables that are highest in pesticide levels, and at least consider purchasing organic produce.[57]

CHAPTER 3

Worldwide, an unknown number of people die from dehydration each year. There is no new sub-species of humans or animals evolving that doesn't need clean water, or needs less water to survive. I have begun to discuss why I think we don't drink enough good water; or conversely, why we drink too much bad water. In this chapter, I will focus on how the human body handles water that is consumed.

Let's start with the condition where someone does not take in enough water. This condition is known as dehydration. It can be an acute/sudden or a chronic/long-term condition. Dehydration is the harmful reduction of the amount of water in the body, according to the Oxford English dictionary. Loss of too much water can cause life threatening symptoms and/or can complicate certain diseases like kidney stones. Newborn human children are approximately eighty percent water.[58] By adulthood, total body water content drops to approximately 60 percent in men and around 50 percent in women.[59] Even severely dehydrated people can be more than fifty percent water, based on a definition of severe dehydration being a loss of 10 percent total body fluid.[60] No matter how dehydrated you are, if you're alive, your body is more than fifty percent water. Therefore, a significant amount of water is needed to stay alive. I believe more is needed in order to thrive and be healthy.

Our internal environment is saltwater. Water serves many functions. It's the transfer medium and the driving force for transfer of substances including electrolytes, nutrients, and hormones, throughout the body. How much water did you transport today? Where did you put it? Can we use it again? We're going to use it again, whether we want to or not, and whether we should or not. Through the physiologic functions of water absorption from the stomach, kidney filtration, and toxin removal by the liver, water balance is maintained. As an ongoing process, these physiologic functions try to remove as many toxins as possible as you re-use the partly new, but mostly recovered and reused water you need. Again, our internal environment is saltwater. Our blood's saltwater is controlled within narrow limits to maintain a constant concentration of salt (sodium chloride) to be 0.9 percent.[61] Either or both too much or too little water or salt in our blood is incompatible with life.

Most water in people is contained within cells (intracellular). About one third of the water left is outside cells (extracellular). About one third of extracellular water is blood, while the remaining volume is interstitial, or water between the cells. Your brain has absolute priority over all biological water rationing and conservation systems within your body. When water or blood is leaving your blood vessels, the first place it may go is into your brain. Imagine your bodies hydration as water filling a wheel with many large spokes. Each spoke goes to a major organ or system of tissues. The arteries going to your brain always want to be full.

The brain's need for water supersedes all other individual organ water requirements. The brain sends instructions to every organ in the body. Water allows this necessary direction and biological signaling to occur. This happens with electrical impulses conducted through fluids and tissues, using hormones affecting water balance, like anti-diuretic hormone, or ADH, and other cellular messengers sending directions to other parts of the body. Most information exchange happens right away. We don't have to think about it. It's happening all the time. It doesn't happen without adequate body water. Dehydration can cause symptoms like confusion, weakness, dizziness, tiredness, and something termed "brain fog" (although not a definable medical term) among other symptoms.[62] If your problem is "brain fog," whatever that is, should I trust you to say you are worse or better?

This reminds me of a sick person I treated in Philadelphia. He was a chronic alcoholic man who was well known to our healthcare team due to his frequent need for admission to the hospital due to acute alcohol intoxication complicated with dehydration. A history was taken at each admission. While taking the patient's history, it was noted that he answered each question asked with an appropriate positive answer. As the healthcare team was leaving his room, he said, “Hey, doc, can you adjust my beach umbrella before you go?” His question was serious. A sterile intravenous solution of water and salt, some sugar, electrolytes, some vitamin B1 or thiamine and a few other vitamins corrected his thiamine deficiency and dehydration. He was able to be discharged and to return to his life.

The brain is eighty percent water.[63] We can’t greatly increase or decrease the water in our brain. Small changes up, hydrocephalus or water on the brain, or down, dehydration, can cause harmful effects on multiple brain functions. The brain is mostly water in a fixed and limited space inside the skull. The brain needs healthy non-toxic water to work right.

When “fight or flight” signals are going from your brain to your body, body water and energy in blood is borrowed from some tissues, like your digestive tract, and sent to others, like the muscles you’re about to use (perhaps the basis for an adult telling you not to swim for thirty to sixty minutes after a meal). When using the fight or flight response, your body experiences a temporary dehydration in the body tissues that aren’t going to help you fight or run away. When you’re about to be eaten, your body is not interested in digesting what you just ate. Your gut deals with food later, if you survive.

After you’re safe, “relative” dehydration shifts your body’s job from escape to protection. Both are needed for life to continue and stay healthy. Your body constantly replaces old cells with new ones. Water is necessary to do so. In my opinion, you need to protect the cells you make. We all know that stress makes everything, including dehydration, worse.

Chronic pain is a stressor and may be an indicator of dehydration.[64] Additionally, dehydration symptoms become fewer and less pronounced as people age. As you age, your sense of thirst lessens and your kidneys aren't able to conserve body water as well. Hunger feelings are often misinterpreted with thirst signals from your brain. You feel hungry but you're actually thirsty. You see body weight is not entirely really your fault, and certainly not your only conscious choice. That last, or first, roll of fat wherever it lands, around your mid-section, on your arms, hanging from your butt, under your chin, or elsewhere, is not all your fault. After age 50, instead of feeling thirsty you may feel tired and sluggish, choosing a nap instead of a glass of water. Prolonged or chronic dehydration can lead to serious complications.[65] Multiple organ systems can have problems from dehydration before the mouth is dry, the blood pressure falls, or there are other noticeable symptoms or signs of dehydration. Eating when thirsty can make people overweight. My advice is, when you feel hungry, drink a glass of water before eating. Drinking water when you're hungry fills your stomach, releases hormones like ghrelin that make you feel full, reduces hunger, and helps you lose weight.[66] Your more fit and maybe thinner body will thank you. Drinking clean non-toxic water is healthier than overeating. Drinking good clean water improves your internal fluid balance. Long standing or chronic dehydration may increase disease risk and shorten your life.

Some caffeine containing beverages and foods have been claimed to either cause or worsen dehydration. These claims were based on one study conducted in 1928. As you can imagine, this subject has only become more complicated over the years. Suffice it to say, the claim that consumption of caffeine containing beverages leads to dehydration has been proven and disproven in rigorous clinical trials over the years.[67] Even decaffeinated beverages contain some amount of caffeine.[68] People may wrongly assume that decaffeinated coffee is 100 percent caffeine-free, when in reality the USDA regulations allow coffee to be labeled decaf if it is 97 percent caffeine-free. Three percent caffeine may seem like a small amount of caffeine, but not if you are caffeine sensitive or consume a large amount of decaf liquids thinking they are completely caffeine-free. The only caffeine free beverages are those starting with no caffeine, like fruit beverages.

Please note I am referring to the consumption of beverages. I am not referring to the use or over consumption of caffeine, or energy drinks containing various amounts of caffeine. Caffeine as a stimulant is a completely different matter and is beyond the scope of this book to explore. The controversy over caffeine consumption remains - good, bad, neutral. Stay tuned. I'm sure this year's study on caffeine consumption will contradict last year's study, but that may be just my skepticism talking.

Your body constantly tries to maintain a temperature of 98.6 degrees Fahrenheit, which is 37 degrees Centigrade, whether you are resting or during exercise. This is achieved through normal water balance supporting sweating and increased blood flow to the skin cooling the body in hot environments. Sweating releases body heat through evaporation. Likewise, increased blood flow to the skin allows for dissipation of heat through the surface of the body.[69] People are skin bags full of salt water. No offense intended.

Exercising in the heat can lead to heat exhaustion when the critical core body temperature reaches about 104 degrees F/40 degrees C, according to researchers. Heat exhaustion can occur in less fit athletes and others when core temperature rises to 101.3 to 102.2 degrees Fahrenheit (38.5 to 39 degrees C).[70] By the time thirst and body temperature are changed by dehydration, the dehydration is already severe. Many essential and critical functions can already be affected or significantly reduced, and you may not feel them or know what is happening. The brain tries to maintain important functions. When you need more water, you have to get more water. Other liquids will help, but only water will fix the problem.

When a person experiences dehydration, circulating blood flow can be decreased by up to a gallon per minute. This decrease in circulating blood volume causes the body to deliver more blood to vital organs and less to the surface of the skin. This makes it harder to cool down because less blood is going to the skin. Compounding the problem, dehydration also decreases your ability to sweat. Even though sweating causes water loss, it is essential to help cool the core body temperature. As a result, replenishing body water allows you to slow the rise in core temperature and maintain body function.[71]

If you're in good health and want to stay healthy, it is recommended that you consume one-half ounce of water for each pound of body weight every day.[72] You may want to drink that amount of water by 5 or 6 pm so you aren't awake peeing all night. Please note, people with chronic diseases like kidney disease or congestive heart failure, as well as other conditions have different water needs. I urge you to check with a competent physician before making any changes in water intake.

Everything fails to some degree without enough water. Water function and balance affects much more than thirst and body temperature regulation. Toxic effects may occur from contaminated water and dehydration. Extreme thirst can change everything you think you know about yourself.

Once you decide on the right amount of daily water you need, it's time to decide how and where to get that water. Don't forget, needs may vary each day and at different ambient temperature.

Municipal water from your tap should be safe according to government guidelines, so the EPA only recommends additional water filtration at home to improve the taste of your drinking water. However, the EPA admits that municipal drinking water can be expected to contain some contaminants, so people with severely weakened immune systems or serious health conditions may benefit from further purifying their water or drinking bottled water.[73]

Many people may choose spring water to maintain their daily minimum water intake. Spring water may be a better choice than some municipal tap water. Spring water is not always as pure as distilled water. Spring water comes from an underground spring and is usually uncontaminated, but it may contain trace amounts of arsenic or other heavy metals. Additionally, spring water usually gets treated before it is bottled and sold. Typically, spring water undergoes some processing and filtering to remove debris and to kill bacteria and other microbial contaminants. Treatment and processing don't remove arsenic or other heavy metals. However, most of the mineral content, such as calcium and magnesium, gives spring water what is described as its "crisp" taste when compared with distilled water.[74] Sodium and potassium may also be present in variable amounts.

Another choice is bottled water. Bottled water is a broad term that can include virtually every type of drinking water. This includes tap water, spring water, filtered water, and ozonated water (water infused with oxygen). Unfiltered tap water has been bottled and sold by some brands and has received criticism over this practice. On the other end of the spectrum, some brands are highly filtered and purified, and some contain additional minerals and/or electrolytes. Some even add fluoride. My intention is not to be a spokesperson for any brand of bottled water - good or bad. I encourage you to read labels carefully or contact the bottling company directly for additional information about any brand you are interested in purchasing.[75]

If you choose bottled water to maintain your minimum water intake, you may be adding to your toxin load and adversely affecting the planet. Most of that water comes in a single use plastic bottle that ends up living forever in a waste dump or dying under water. I will focus on the topic of plastic in the environment and its effect on water in a subsequent chapter, so please stay tuned.

Purified water is another choice but may be confusing because it is so general. Purified is a term used to describe water that has been "purged" in some way. Filtration is minimal physical purge. Filtration can be accomplished via an activated charcoal/carbon filter, think of those countertop pitchers or dispensers. It can be accomplished with a ceramic filter. Many homes have a reverse osmosis system built in for water filtration. Water purified by reverse osmosis is the most similar to distilled water in terms of taste and mineral content.[76]

Alkaline waters have come into vogue recently. Actually, keeping the body alkaline is not a new concept. Understanding alkalinity requires a basic understanding of pH. pH is a numerical scale that is used to describe how acidic or alkaline a solution is. In the human body, the arterial pH stays very close to neutral and is tightly controlled biologically. pH affects many functions in the body, and one of the most important is the delivery of oxygen from the blood to all body parts, tissues, and cells. Extreme variations in pH are not tolerated and can cause death.

In 1931 Otto Warburg won the Nobel Prize for experimenting with body pH and cancer. He discovered cancer cells only grow in acidic pH. Thus, began the move toward making people's bodies and body water more alkaline or "basic." The best way to make your body alkaline is to minimize or avoid eating and drinking acidic substances like coffee, meats, fish, and sodas. You don't need to consume special alkaline substances, or alkaline waters, just avoid most acids.

After ninety years, we still don't have proof of this approaches long-term effectiveness and personal outcomes. The incidence of cancer is still rising. There are books (e.g., Robert O. Young and Shelley Redford Young's *The pH Miracle, Balance Your Diet, Reclaim Your Health*) to help you avoid acidosis or acid pH.[77]

You may go through your entire lifetime without ever knowing your arterial blood pH, wanting to understand more about it, or needing to have it checked. Most healthy people do. These few paragraphs about arterial blood pH are already too much to learn and understand if you have no use for them.

Cleaning your body water and restoring fluid balance, or hydration, helps people survive. Leaving body water toxic and dirty isn't a healthy or good prolonged choice. To change, we don't have to learn new things, we have to unlearn destructive, older, automatic behaviors. Doing that will reduce symptoms associated with contaminants. There is also a step two. If you add necessary nutrients to your diet, your overall health outcomes will improve even more. Throw in some exercise and health improvement can be maximized and maintained.

I believe time is limited and fixed for individuals and the human species.

What is a keystone species? A keystone species is a species on which other species in an ecosystem largely depend, such that if it were removed the ecosystem would change drastically and possibility irrevocably. Humans are the super keystone species. Humans drive everything, all change, the only constant variable. Time is not limited for the water or the Earth. Individual troubles, natural and manmade, threaten the integrity and performance of Earth's functional life supporting systems, especially the water. There is little non-toxic "natural" water remaining on planet Earth. No known current peril, other than people, menaces the water and planet's survival. The Earth has been here much longer than people. It will be here long after humans are gone.

Get ready for forever.

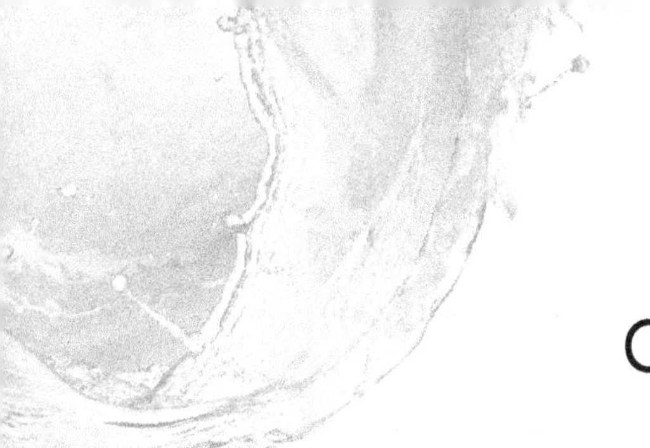

CHAPTER 4

The first step to doing things right is to stop doing what you are doing wrong. The longer you delay by doing wrong things, the more time it will take to do the right thing. People are dying from dehydration right now. Fix everything so there will minimally be a different better later. But fix water now or there will be no later for humans. We aren't close to total water failure at this moment, but we are closer than we have ever been.

People should be much kinder to all forms of water.

Have I convinced you, or confirmed your realization that staying hydrated is important? Let us move on to how to stay hydrated with the cleanest, purist water available. My choice for daily hydration, or as needed rehydration is to drink distilled water. Distilled water can be easily made at home or can be purchased at your local grocery or convenience store. I recommend you purchase distilled water in glass containers rather than in plastic ones. Everywhere I've lived in the United States I have been able to get distilled water in glass containers. Please try and do the same.

Why is distilled water the purist best water source? Distilled water is made from pure oxygen and hydrogen gases. Two gases are put together under the correct circumstances and make liquid water. Those gases exist only in pure uncontaminated form. They are condensed to form pure unadulterated molecules of water. Home distillation units can be easily found online, take up little space, and inexpensively produce enough distilled water daily for your drinking and cooking needs. Except for distilled water, all water exposures have the potential to introduce toxins into your body. In my opinion, only distilled water is pure and safe.

On a personal note, I regularly bought distilled water in glass containers at a local health food store. Trying to be an informed consumer and with the knowledge and interest I have in obtaining clean water, I asked about the distillation process used for the water I was buying to drink. I was told the water went through a reverse osmosis filter and was then distilled. The product label listed these processes. The water labeled as "distilled" was produced into, stored, and sold, in glass bottles.

Let me take a moment to describe these processes. Reverse osmosis is a means whereby pressure is applied to tap water causing the water to pass through a membrane filter to purify the water, leaving the contaminants on one side, and only pure water on the other.[78] Some of you may have reverse osmosis or R/O systems in your home. I do. I also distill my drinking water after producing it with an R/O system.

After R/O production, according to the product label, the water in the health food store was then distilled. This means that the water has been boiled to become steam that exists only at two hundred twelve degrees Fahrenheit, the same as one hundred degrees Centigrade. The steam was then cooled to condense gaseous hydrogen and oxygen into liquid water again. Then it is totally free of minerals and salts.[79] Distilled water should be produced into and stored in clean untreated glass bottles to remain free of added substances.

Besides reading the bottle's label, I met the person who produced and distributed the distilled water that was sold in several stores. We began a discussion about what type of distiller and distillation process was used. The person who prepared the water told me the water was not actually distilled (NOT DISTILLED). I asked again, not hiding my surprise or anger. The product started as tap water, like most every other commercial liquid for sale. The water was bottled after a reverse osmosis filtration process alone. This is not the same as steam distillation despite what anyone miss-states. The label stated the water was distilled when it was not.

The product was also sold in a dozen different local independent health food stores. The minimalist lesson here is you need to thoroughly read every label. Ask questions. Sometimes, even that is not enough to get information you need to make informed decisions about water your family uses for everything. You hear, see, read, and otherwise access information about some of those labelled products being withdrawn from the United States and other worldwide markets. Sometimes the product recall is rather sudden and hurried. You may need to go further to get reliable information about the products you buy for yourself and your family. Pet food problems have recently been added to my "check everything before purchase" checklist. It is a long road, but could contaminants from anywhere worsen pollution anywhere else? I believe part of the problem may be that claims made may not be reviewed by regulating agencies unless there are multiple reports of illnesses or deaths possibly associated with a particular product.

Make change when you get trustworthy information. Adjust thoughts and behaviors as suspicion increases before it's officially recommended to do so. Stay ahead of the continually worsening water change curve.

In many personal communications with allopathic, osteopathic, chiropractic, homeopathic, and naturopathic physicians, shamans, and others, many suggest drinking distilled water when possible, and often recommend it to their patients. They do that to avoid or minimize accumulation of toxic materials in patients. These drinking water selections have no large objective test evaluations either proving or disproving those subjective health claims. Your ultimate and ongoing series of personal choices should be, and always are, up to you. Be objective about how you feel. Be as safe as you can be knowing what you know and don't know. Grow your water knowledge base whenever and wherever you can. Allow valid information to protect you.

I believe distilled water is the optimal safe drinking water choice. I recognize there are many smart people thinking distilled water may not be the best drinking water option. Those people have many good reasons for thinking the way they do. What about phase 4 water? Structured water? Alkaline water? Minerals in water? The list of inquiries goes on. People's opinions are difficult to prove. Those assessments are also difficult to disprove.

I focus on how I feel both at minimal activity through and during my heaviest workout energy expenditure. Don't be confused by people telling you how you should feel. Stay with what works for you.

Distilled water containing toxic metals is not possible unless contamination happened post distillation. Distilled water is uncontaminated and pure.

I don't believe the "opinions" about distilled water washing minerals out of your body.[80] It would be an easy study to do. I disagree with those opinions and feel they don't present objective proof for their claims. I can find more references that support my opinion than those touting possible dangers of drinking distilled water.[81] Minerals aren't leeched from bones by water. All water goes through changes and tissues before the water-based substance called "blood" contacts bones. Plain, tap, mineral, demineralized, ionized, distilled, alkaline, or other waters never contact bones directly. No free water contacts bones in humans. Mineral deficiencies are caused by not eating a variety of mineral containing foods.[82]

Another acceptable drinking water choice is water filtered by reverse osmosis sometimes referred to as R/O water. In some areas of the country with "hard" water, or water with a high calcium content, homes either come with built-in R/O systems or these systems can be added after construction. In any case, reverse osmosis filtration offers a good alternative. No physical filter accomplishes total initial toxin removal and most filters become less effective over time, even relatively short times.

Unless being told to change water filters daily, the manufacturers schedule to change filters is always longer than it should be to maintain the cleanest water. Filter contamination happens when the first filtration occurs. The worse your water quality is, the sooner and more often water filters get contaminated. Filter contamination worsens more with every subsequent use.

This easy in-home choice for water purification is my best recommendation for daily water use and consumption. Like many recommendations, I expect them to change over time. There are other in-home filtration devices that either attach to your faucet, filter water into a container, or are part of your refrigerator water dispensing system. Although these filters help in removing some contaminants, it's up to the end user to change the filters frequently and use them properly to get maximal benefit. In my opinion, these devices are an improvement over tap water but are not as good as either distilled or R/O water.

As you correctly surmise, I am not a fan of drinking tap water from many, most, or maybe all municipal water supplies. I held this opinion long before the change in Flint, Michigan's drinking water from Lake Huron and the Detroit River to a source from the contaminated Flint river. That change resulted in lead and other heavy metal toxicity for numerous Flint residents, including children. The drinking water crisis reported from Flint, Michigan in 2014 was declared a federal state of emergency in 2016. The problem was expected to be resolved in 2019, but as of 2022 it has not been. In my opinion, most public tap water is unhealthy. Even good water may go through bad toxic piping. When bad water is supplied with bad piping the problems become magnified.

Water similar to Flint's poor quality can be found in hundreds of other local water supplies in this country.

According to the CDC, "What comes from your faucet is not water ... it's a toxic soup of chemicals, bacteria, viruses, and metals." Please read the following explanation. Although I have read this statement or attended lectures with speakers stating this information numerous times, I am unable to find a trusted primary published reference for this information. Did the CDC really "say," or print this testimony for public examination? This information could be wrong without anyone knowing. Should we believe it or not? I have not found objective believable proof for that statement. I am still looking.

The government admits to this. What don't they know? What don't we know? What they tell us is both good and occasionally bad enough. Chemicals, bacteria, viruses, and metals can make you sick, may shorten your life, and can kill you. One careless person, one wrong decision, or one toxic substance can harm or kill people. We saw this play out in Flint, Michigan. One person working to clean the water can't repair this toxic water situation. Properly informed, the right person may be able to keep you safe.

Do you know where your tap water comes from? For most of us, it comes from either surface water or ground water. Surface water means water collects in streams, rivers, lake, and reservoirs. Ground water collects in pores and spaces within rocks and in underground aquifers. Ground water is obtained by drilling wells and pumping the water to the surface.[83]

Public water supplies come from both surface and ground water. Treatment systems are either run by the government or privately held facilities. If surface water is used, it is drawn from the source, treated, and then sold or delivered to your home. Ground water systems do not always treat the water before selling it or delivering it to your home.[84]

If you have a private well, ground water is the source of your water. Owners of private wells are responsible for ensuring their water is safe from contaminants and toxins.[85]

Since the responsibility lies with the individual, I cannot comment on the purity of the water sourced from private wells. I would, however, recommend that you investigate nearby sources of potential contamination such as agricultural areas, waste sites, fracking, and/or nuclear facilities to name a few.

Too many people in my opinion, choose bottled water as their preferred drinking water source. Let's focus on the water going into these bottles. In the next chapter, we'll focus on the bottle.

Some common sellers of bottled water, Pepsi®'s Aquafina® and CocaCola®'s Dasani® start with tap water.[86] Not all bottlers tell you that. Their processing doesn't have to remove toxins. Bottled drinking water also usually starts with tap water.

Each time people try to enhance or change water they often leave some substance behind, add other substances, or both. Flavored, vitamin, mineral, herb infused, oxygenated, ionized, sparkling, alkaline, dream, and magic water are all waters altered by people. Many products started as tap water. It is my opinion many of these "enhanced" water products try to add a feeling of good health without proper controlled studies or information supporting, documenting, or proving their claims. Some waters labeled as specially made or new waters aren't improvements in my opinion. Except for added components, new waters have the same actions as old waters. Every new water starts from aged water. Advertisers want you to believe new waters are better for you. Advertising about some water benefits is brainwashing persuasion designed to separate you from money. Otherwise, water could still be provided free of charge.

The best advice I can offer is to ignore unsupported water "labeling schemes." Many of these water products are owned and produced by some of the world's largest for-profit companies. Corporate goals are to generate profit and keep owners and shareholders satisfied and pleased, not necessarily to produce good health in the users. Nor is there a reliable measure of good health in people. There are many patho-physiological indicators of bad human health. People eat and drink with their eyes first. If something looks appealing, we are more likely to eat or drink it. I would advise to not drink liquids made with artificial colors or non-natural flavors. It doesn't matter how tasty or pretty these things are if they could be potentially harmful when you drink or eat them.

Even flavors labelled "natural" may start as a natural product but are enhanced in a laboratory, and do not come directly from a plant source.[87] The word natural doesn't have a consistent or dependable definition.[88] Some people might define natural to mean anything that appears on Earth right now. Some people say, "of course it's natural, it's here on the Earth, isn't it?" In my opinion, water is perfect in its most pure form, and nothing done to it has ever made it better. Distillation removes toxins, natural ones like arsenic and others discarded into water. I hope we can all agree added substances, whether natural or not, may worsen water quality.

What about micro-dosing pure water? What about using micro-dosing to expand the total body volume of liquids, just a little bit, perhaps only for a short time? That is what drinking anything, or eating high percentage water content fruits and vegetables, initially does. Increasing total body water volume reduces the concentration of toxins to be eliminated and eases the stress on internal systems continually working at detoxification processes. Endogenous internal chemicals and neurotransmitters are made within humans.

Water is essential to everything going on that is human. No one knows what or where the mind structure is, but I think there is water there with it helping the mind's workings.

This next example is for the kids reading this book. Hello, young humans. I am sorry to have to scare adults into action. I know you deal with dwindling natural resources daily. That affects us all. The tides affecting seventy percent of planet Earth may be covering the homes on the shores sooner than we think.

There are many other products called "all natural" by the manufacturer. Don't be fooled by labeling on anything that reads "all natural." The word "natural" has no consistent meaning as mentioned above. It seems to me that even water isn't "all natural" now. Natural doesn't always mean good. Synthetic doesn't always mean bad. Evaluate both information sources before you believe or use them.

Water's actions don't vary with new and "improved" waters which aren't cleaner than distilled water. As mentioned previously, some of these waters are unfiltered bottled tap water. My opinion is filtered water isn't necessarily better than tap water Even when filtered, filtration does not remove all foreign materials. Some of these waters have an additional side effect of making your purse lighter or wallet thinner due to their high cost. The water may be packaged in plastic, potentially adding additional toxins, plasticizers, phthalates, and BPA. These toxins will be discussed in detail in a later chapter.

I have found no objective reproducible proof these waters provide better health. I suggest you review all information before you believe it or drink the water of your choice.

Water research should be conducted like any good research, using the scientific method. This process includes the following steps: ask a question/identify a problem, conduct background research, form a hypothesis, experiment and observe, then draw a conclusion based on the findings.[89] Research should be conducted on subjects/people with various abilities, inabilities, diseases, and physiological challenges before any sound conclusions can be drawn about the population using that water. Research should be published or made available to experts for review and evaluation.

If you're eating, you need to balance water intake with the water from the fruits and vegetables you consume.

Some fruits and vegetables are more than ninety percent water.

These foods include fruits such as watermelon, cantaloupe, grapefruit, and others. High water content vegetables include broccoli, cauliflower, celery, cucumber, and others. These fruits and vegetables count as water sources and should be used to figure water balance.[90] Consider buying organic fruits and vegetables having high water content to minimize possible exposure to toxins. In my opinion, this is a wise use of your money. Places to save money is on fruits and vegetables that you peel. Don't pay extra for things like bananas or avocados.

This leads us to another water source, juice. Juice from a fruit or vegetable may be better hydrating than plain water because it contains active nutrients and mineral salts getting absorbed with the juice.[91] These substances include sugars or carbohydrates, essential fats, amino acids, proteins, vitamins, minerals, electrolytes, and other nutrients.

Look for "organic" non-concentrated juice sold in a glass container. Avoid juices made from concentrates. Read the label to determine if the juice concentrates are made with sugar from genetically modified sugar beets or other sweeteners. Check to see how the fruit or vegetable was grown. Verify how the juice was produced. If you've reason to not believe the seller or they have a history of false statements or products withdrawn from the market, protect yourself. I offer the same caution with other forms of non-cows' "milk," like rice or nut milks. These products rely on tap water too. If possible, make your own special milks with organic rice or nuts, and distilled water.

It's difficult to eat healthy while drinking unclean water. Possibly contaminated water is used to grow food. More than fifty years ago, Rachel Carson wrote all water is contaminated.[92] I hope you all can agree with my opinion most tap water has become more polluted every day since.

Last, and certainly least in my opinion is soda, pop, soft drinks, whatever it's called in your region of the country. Either avoid or limit your consumption of these drinks. As previously discussed, the main manufacturers and bottlers of these products start with tap water. To this impure water source many other ingredients may be added such as high fructose corn syrup (HFCS), artificial sweeteners, artificial colors, etc. Then adding insult to injury, the resulting beverage is put into a plastic container. This is not your best choice for hydration.

As a healthcare clinician, I spend my professional career working to manage water balance and care for patient's health. If you have your health, you can redefine your everything. This information should be used to change the water in your life, everything from drizzle and drops, to downpours. We're all at fault. We all share the guilt. We all should strive to be part of the solution. If not, the solution will always be incomplete.

Words on these pages are a call to action to help address the water inside us, around us, falling from the sky, and the water we drink. In my opinion, bad water is making us sick. According to the WHO the number of deaths from water related diseases, including infections, exceeds 5 million people annually.[93] I think water issues should concern every person in every country. I fear water effects and water politics may end life as you know it. Have they killed the future you've imagined? Imagination, whether good or bad, is the preview of coming attractions and approaching distractions. The water lies you drink every day may be affecting your health in ways you don't realize.

Toxins in water poison drinking water, the Earth, oceans, sky, soil, food, animals, people, and other life forms. My opinion is our current drinking water choices are awful. People die from drinking water.[94] I would like all of us to demand change to fix the water however we can. Water is essential to every living organism on Earth. I believe nothing is more important. Although the stream ahead seems strewn with rocks, I think of them as steppingstones, not stumbling blocks. Go forward to reclaim the clean waters of the past. I believe many people want that and everyone will benefit.

I believe nothing is more important than restoring and maintaining good health and wellbeing. Drinking non-toxic water helps accomplish those goals. Some average weight people need to drink eight eight-ounce glasses of water a day to be healthy; others do not.[95] The volume of water needed for good health is a variable range, not an absolute number. Daily water intake and needs vary with body type, weight, exercise, water intake, food type, and food intake.

Those things change throughout the day. They may change with body temperature and the surrounding areas environmental temperatures to which you are exposed.

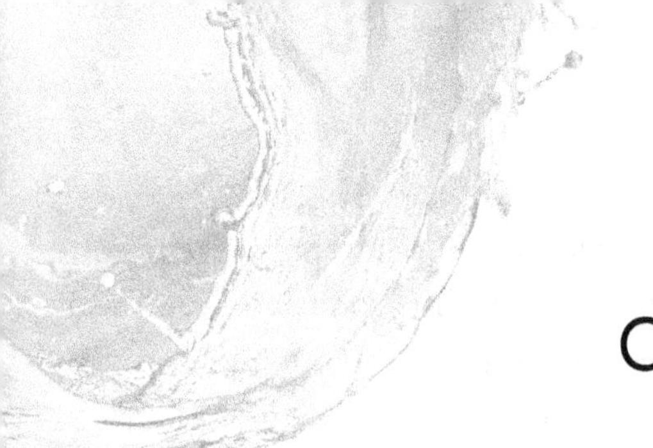

CHAPTER 5

The water supply and quality must be sustainable for life to be maintained. If clean, pure water does not survive, humans will not either. I do not want to preserve any current aspect of water here on planet Earth. It all needs to be fixed first. Fix it. Don't just improve it. Work on it. Make it right again.

Just as important as the water you drink, your drinking container must be examined and evaluated. Once upon a time when the water was clean, we could cup our hands, get water from the freshwater source, and drink freely from the container our hands made. Those days are gone. This chapter will discuss decisions you make about the drinking vessel, water, beverages, and other liquids you consume daily.

Drinking liquids from clean, non toxic, untreated, unpainted, glass should be a priority for everyone, especially babies, when possible, toddlers, young kids, and elderly people. There are ways to do that correctly at low or no risk. I am concerned about exposure to toxins from drinking vessels. Let me tell you why.

If you drink from synthetic foam or polystyrene containers you may be undoing the good work you did choosing the cleanest, purist water you put into that container. Synthetic foams and polystyrene are made with chemicals known to be probable cancer-causing agents in people.[96] Styrene (vinyl benzene) is a chemical used in the synthesis and manufacture of polystyrene, as well as hundreds of different substances people contact every day. Almost all industrial resins produced from styrene can be grouped into six major categories, with polystyrene making up half of those resins.[97] In 2006, the US produced 11.4 million pounds of styrene.[98]

I try to reject using all probable cancer-causing substances. I hope you will too. Not only might these containers affect the liquids you put into them, but these substances are also very hard to remove from the environment.[99] We all know some synthetic foam and polystyrene containers are being banned from many states for "to-go" containers.[100,101] The consequences of your decision to use, or not use, polystyrene containers can have serious long-lasting affects on your health. They can have sustained effects on the environment.[102] A global look at the personal container choices will be covered in a forthcoming chapter.

I hope you have already made the personal choice to move away from synthetic foam and polystyrene containers. However, those containers are not the only ones of concern in my view. Certain plastics also can create situations undoing your other good choices.

Since the 1960s, Bis-Phenol A (BPA) has been used in food packaging and beverage bottles. Some plastic cups, food containers, bottles, and metal cans are made from and/or coated with BPA. When food comes into direct contact with any packaging material, small and measurable amounts of the packaging may migrate into the food or liquids and ultimately be consumed entering the human consumer. Review of the safety of food packaging materials is the responsibility of the Food and Drug Administration (FDA). The FDA assesses the likelihood of packaging materials migrating into the food chain to assure that the movement of these agents stays at safe levels.[103]

Recently, there has been a call from the public to remove BPA from certain containers used for infants and children. As a result, in 2012, the FDA banned the use of certain BPA-based materials in baby bottles, sippy cups, and infant formula packaging.[104] Due to heightened interest in the safety of BPA in food containers, scientific studies have been conducted to address concerns ingesting small amounts of BPA may be harmful. The FDA and National Toxicity Program, partnering with the FDA's National Center for Toxicology Research, carried out in-depth studies to try to answer these questions about BPA safety.[105]

As of 2014, the FDA has concluded based on the most recent safety assessment, BPA is safe at the current levels occurring in foods and the safety of BPA is currently approved for uses in food containers and packaging.[106]

Although the FDA's position is that the levels of BPA absorbed are safe. BPA can leach into food from many consumer products like polycarbonate tableware, food storage containers, water bottles, beverage bottles, etc. The amount of BPA that leaches into liquid from a polycarbonate bottle depends on the temperature of the liquid in the bottle.[107]

If you put pure water or organic food into plastic, are they still pure and organic? There's no such thing as organic plastic. Food wasn't organic when you bought it in plastic and when it was stored in plastic. You may buy organic pasta to be boiled in the plastic bag it comes in. Heating food in plastic increases the leaching of plasticizers into the food you eat. Those chemicals pass into your body from eating the "organic" food. Cooking food in a plastic bag may be a quick way to prepare a meal, but it can create a potentially serious exposure to toxic substances. Those chemicals enter your body and some pass into the water table through urination. These toxins can possibly harm you and your family as they cycle through the water and food systems.

Exposure to BPA is widespread. A study was conducted by the National Health and Nutrition Examination Survey (NHANES) and Centers for Disease Control and Prevention (CDC). The NHANES III study conducted in 2003-2004, found that 93 percent of 2517 urine samples from people aged 6 years old and older in the US had detectable levels of BPA. BPA can also be found in breast milk.[108]

Some animal studies suggest infants and children may be the most vulnerable to the effects of BPA.[109] Perhaps not surprising, a 2009 study found 232 toxic chemicals, including BPA, in umbilical cord blood of newborns.[110] I believe that many more substances could have been found if they had been looked for. This data convinces me that newborn babies start life exposed to numerous foreign substances. When infants go out into the world, they can be bombarded with thousands more chemicals.

It's not just infants and children we should be concerned about BPA exposure. BPA is an endocrine or hormonal disrupter. Animal studies have demonstrated the mechanisms through which endocrine disruptors influence the endocrine system to alter hormonal functions. Endocrine disruptors can have several different actions in the body. They can mimic or partially mimic naturally occurring hormones like estrogens or female hormones, androgens or male hormones, and thyroid hormones affecting all humans, and can potentially produce overstimulation. They can bind to receptors on cells and block the action of the hormone potentially decreasing the hormone action. Or the disruptors can interfere or block the way natural hormones are made or eliminated by the liver.[111]

This is important because even very small changes in hormone levels can alter the body's endocrine signaling system. In 2000, the National Institute of Environmental Health Services (NIEHS) and the National Toxicology Program (NTP) convened a panel of experts founding there was "credible evidence" some hormone-like chemicals can affect test animals' bodily functions at very low levels, well below the "no effect" levels previously determined by traditional testing. BPA is reported to have estrogen or female hormone like effects in both women and men.[112] I don›t believe that people should be exposed to these potential effects without their knowledge, permission, or active choice.

Hormone disrupters have been associated with fertility issues and cancers in people. "Associated with" isn't the same as "caused by." Why take the chance if you don't have to? Following are some effects hormone disruptors can have in both men and women.

BPA is a weak synthetic estrogen-like substance. Estrogen can make hormone-receptor positive breast cancer develop and grow, so many women choose to minimize their exposure to BPA. Additionally, a 2011 study found that pregnant women who had high levels of BPA in their urine were more likely to have daughters who showed signs of depression, anxiety, and hyperactivity seen as early as age three.[113] Boys aren't affected in the same way for reasons not clear.

Plastics are hormone (endocrine) disrupters. Some plastics can increase endocrine effects, some can decrease it, and some could make normalcy, whatever that really means rarely happen. Most people are referring to averages. Everything secreted by a gland is a hormone. Hormone crashes cause health problems. No one thinks hormone failure effects are beneficial or good for people, drinking water, or reproducing by achieving a pregnancy. In my opinion, there are minimally hundreds, and probably thousands, of hormone disruption substances in water. Some hormone disruption effects are thought to be causes of non-fatal and some fatal hormone cancers. Thyroid disease is an epidemic in the United States. What role do hormone disrupters in drinking water have? The definitive answer is unknown. Is increasing harmful chemical contamination making these problems worse, better, both, or neither?

When drinking water contains hormone disrupters, GMO's, pesticides, or herbicides, that water is potentially harmful or deadly to people.

If you didn't plan ahead and store water prior to a hurricane, what are you drinking after the storm?

A 2014 study conducted by NIEHS found preliminary evidence of an association between BPA exposure and the occurrence of prostate cancer in men.[114,115] Like breast cancer, high levels of BPA were associated with low levels of the prostate cancer marker prostate specific antigen (PSA) in patients under 65, which may create a false negative.[116,117] The study was published in March of 2014, and found a significant difference in the levels of BPA in the urine of 27 men with prostate cancer versus 33 men without prostate cancer. The primary researcher in the study postulated the removal of the toxic agent might be an effective strategy in reducing the risk of prostate cancer.[118]

Should we believe if BPA is an estrogen-like substance and has a potential effect on male sex hormones (androgens), it may also affect the onset of puberty in both males and females. The evidence is more compelling in females experiencing early puberty than in males.[119]

Type 2 Diabetes Mellitus has also been attributed to BPA exposure, but the data are preliminary and conflicting due to study design and varying exposures to BPA.[120] In any event, knowing that BPA can mimic other hormones, it is conceivable to me BPA can affect the pancreatic beta cells producing insulin and causing dysfunction or their premature death resulting in relative insulin insufficiency and diabetes.[121]

Hand in hand with type 2 diabetes, is weight gain and obesity. In a study performed by the Canadian Health Services, a Canadian Health Measures Survey of 4733 adults aged 18 to 79 was conducted to examine the relationship between BPA exposure and increased body mass index (BMI). Higher urinary BPA levels were positively associated with a BMI of obesity. Higher urinary concentrations of BPA were associated with more weight gain. This study contributes to the growing body of evidence that BPA is positively associated with obesity. Additional prospective studies with repeated measures are needed to address temporality and to improve exposure classification.[122]

There are also some data on the effects that BPA may have on behavioral problems in children. In a study conducted by the Department of Environmental Health Services that included 250 children (135 girls and 115 boys), it was found that exposure to BPA in the prenatal period and in early childhood may affect behavioral outcomes in sex-specific and timing-specific manners in inner-city children. The results of this study raise concern about the widespread prevalence of this chemical and its potential effects in children from early-life exposures. In the study, the direction of associations seen between BPA concentration and behavioral symptom scores depended on the timing of exposure and differed for boys and girls. It's clear to me the effects of BPA in relation to human health are complex multiple symptom cluster presentations and multi-factorial ever changing patho-physiological causes. Further research in both humans and animal models is recommended to explain and detail the effects of BPA on brain development and behavioral outcomes, especially in children.[123]

Sometimes the chemical Bis-Phenol S (BPS) is substituted for BPA. BPS became a replacement for BPA because it was believed to be less leaching than BPA. The hope is if BPS was less leaching than BPA, people would consume or absorb less of the chemical.

Despite this hope, almost 81 percent of Americans have detectable levels of BPS in their urine.[124] Once BPS enters the body it can affect cells in the same ways that parallel BPA. In a 2013 study conducted by The University of Texas Medical Branch at Galveston, it was found that even minute concentrations lower than one part per trillion of BPS can disrupt a cell's normal functioning. That could potentially lead to metabolic disorders such as diabetes and obesity, or even cancer. Even if you see the words "BPA-free" on the label, which is true, the manufacturer may neglect to inform you that they've substituted BPS for BPA. BPS has not been tested for the same kinds of problems BPA has been shown to cause.[125] In my mind, the toxicity of BPS is unknown. Some investigators think BPS is as toxic as BPA. Those chemical toxins get into people from the water you drink and from some containers you drink from.

This is what we know about BPA and its relative, BPS. Do you want to take chances on what we don't know? Most people take those chances daily, at every meal, and between meals. Neither safety nor toxicity of BPA or BPS have been objectively proven to my satisfaction. I choose to act conservatively by avoiding products packaged in containers using BPA until it is proven to be safe. That proof is still forthcoming in my opinion.

BPA is the perfect hidden synthetic compound. You can't see, smell, or taste it. You can't tell it's there from touch. Both the CocaCola® and PepsiCo® companies market water and other beverages in plastic bottles. Do their plastic bottles contain plasticizers or BPA? Most labels don't read yes or no.

Does BPA or BPS, alone, in duo, or in other combinations hurt people or the Earth? It could be either or neither. What if it does both? As previously mentioned, I try to limit my exposure to potentially harmful substances. As a parent and a caregiver to a ninety-five-year-old relative, I make the personal choice to reduce exposures of my family to BPA. I think the best way to start detoxing a baby or kid is to first stop them from collecting more toxins. Stop bad behaviors before you start teaching kids what good things they should do. Stop kids by stopping yourself. Kids do what they see. They may automatically download the behaviors seen in adults. Be a good example for their eyes and actions.

What can you do to limit your exposure to BPA? I suggest the following strategies to limit exposure to BPA and/or BPS:[126,127]

- Don't microwave polycarbonate plastic food containers. Polycarbonate is strong and durable, but over time it may break down from overuse at high temperatures.
- Use glass, porcelain, enamel-covered metal, or stainless-steel pots, pans, and containers for food and beverages whenever possible, especially if the food or drink is hot.
- Plastic containers have recycle codes on the bottom. Some, but not all plastics are marked with recycle codes 3 or 7 may be made with BPA.
- Reduce your use of canned foods.
- Use baby bottles that are BPA free.
- Carry your own glass, steel, or ceramic water bottle filled with distilled or R/O water.
- Plastics with recycling symbol 2, 4, and 5 are generally considered OK to use.
- Plastics with recycling symbol 7 are OK to use as long as they also say "PLA" or have a leaf symbol

on them. The recycling symbol number is the code that shows what type of plastic was used to make the product.

- Recycling symbol 1 is also OK to use, but shouldn't be used more than once (no refilling those store-bought water bottles!). Keep all plastic containers out of the heat and sun.

What more can you do to enjoy and consume pure, clean water? My best recommendations are to store and drink it in non-reactive containers. Examples of these containers are unpainted glass, non-reactive silicon, stainless steel, enamel, or porcelain. Both the Earth and I implore you to use reusable, non-reactive containers for your health and for the health of the planet. Americans use more than two million plastic bottles every hour, that's almost fifty million bottles used every day.[128] Where does all that plastic trash go? Much of it goes into the water! BPA contaminates most drinking water systems. In the next chapter, we will examine the consequences of dumping plastic into the ocean.

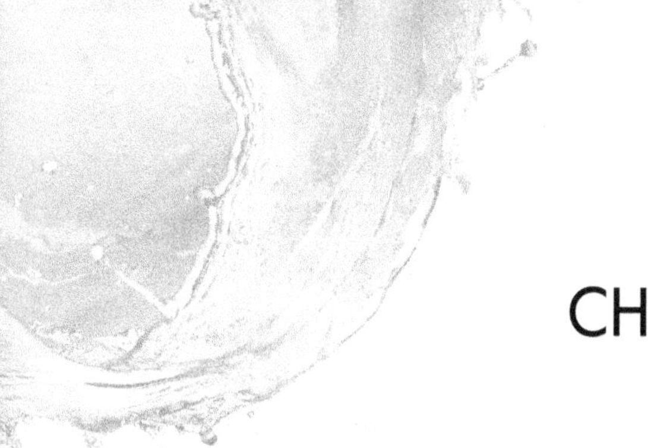

CHAPTER 6

In the 1967 movie *The Graduate,* Benjamin Braddock was told by Mr. McGuire that there was a great future in plastics.[129] That prophecy has been proven true. Plastic is everywhere now. Plastic trash is everywhere too, everywhere it shouldn't be, especially in water.

Mass production of plastics began seven decades ago, in the 1950s. Since then, plastic manufacturing has accelerated so rapidly and 8.3 billion metric tons of plastic has been created. Not only have we made a lot of plastic, but we also keep making more and more year after year.[130] Plastic was made to last. And last. Plastic lasts longer than the current expected human lifespan of three score and ten years plus or minus who knows how much expected variation to help define expectations for normal, average, elderly, and more.

"The people of Earth, roughly one hundred eighty countries containing people and water, out of Earth's two hundred or so countries, are being assaulted by another killer virus pandemic, we think. The water is also being violated by toxins made by people and discarded into water. Are the viruses making that worse? Is water either directly or by some sort of secondary intent always getting worse?"

That is two strikes out of three possible already. Is it too late for any changes to have an affect on what seems inevitable? Is it too late for water and everything that flows into, because of and from it?

Billions of plastic water bottles reading Dasani®, Aquafina®, Nestle®, and other names overfill the landfills on this planet. Nestle® markets water under a half dozen different brand names.[131] Some companies are making progress and reducing plastic litter. Some needs to become most, most needs to evolve into all, all needs to grow and materialize into everything, now! Now must become the present time before it is too late to change the future.

It's not just the companies bottling beverages in plastics needing to change. All companies using plastics and consumers of single use plastic items need to deal with the waste we produce. McDonald's has announced by 2025 all of their packaging will be made from recycled plastic.[132] Plastic manufacturers need to develop biodegradable polymers.

Amazingly, approximately 60 percent of all the plastics ever manufactured are still on the planet somewhere.[133] Landfills and the ocean are the final destinations for three quarters of non-biodegradable plastic trash. These statistics support the notion that we are not very good at managing plastic waste. In 2017, it was reported in a Science Advances article there is currently 4.9 billion metric tons of plastic debris in the environment. This amount will more than double by 2050.[134]

Plastic waste has three fates. It can be recycled; it can be burned; it can be sent to landfills.[135] Some of these solutions seem better than others. It's worth taking a minute to examine each option.

The cumulative plastic waste generated since 1950 has reached 6.3 billion metric tons. Twelve percent has been incinerated and nine percent has been recycled.[136] That leaves more than seventy-five percent ending up in the environment. Looking at just the US, less than 10 percent of plastic waste is recycled. We'll look more into recycling plastic and its fate later in this chapter.

If plastic is burned, it is eradicated. But it's not that simple. If it's burned at low temperatures, which often happens in third world countries, noxious fumes are released creating a health risk. In developed countries, environmental consequences depend on the incinerator and emission technology.[137]

A significant percentage of trash in United States landfills is plastic.[138] Much of that plastic spills into water everywhere. Everything thrown away on a planet mostly made up of water eventually pollutes the water. Let's follow the path of everyday plastic disposed of as household waste. You discard your single use plastic water bottle and it goes to the landfill. There, the plastic bottle is combined with other waste and is buried under soil. Plastics entering landfills are said to take 500 years to breakdown.[139]

Is this true? The model reporting a 500-year degradation time is just that, a model. Models differ and change.

We cannot know for sure. Plastic hasn't been in use for 100 years yet. What is known is plastic trash in water will last many more generations. As plastic continues to be made and used, water will continue to become saturated with waste, both macro and micro plastic scraps. Tiny un-seeable micro plastic particles already contaminate some water sources, including some drinking water. Plastic lifespan is both its appeal and our dilemma. What will future archeologists think about our civilization as they sift through tons of plastic pollution in water everywhere, and in waste disposal sites? In my thoughts, will plastic be found in future samples or biopsies of human tissues?

Think about all the plastic you contact each day. I tried not to touch anything plastic for one day. I couldn't do it and challenge you to try the same. From your toothbrush in the morning to the K-cup® you use to brew your coffee or tea. A trip to Starbucks®? More plastic. Television remote control to catch the news. More plastic. You see my point.

Plastic buried under soil in landfills can enter our water supply, especially if the landfill is older than 1990, as most are. Prior to 1990, landfills were not required to be "lined" and have the potential to leech toxic chemicals into nearby groundwater sources that can enter the drinking water system. After 1990, landfills were required to be lined with two feet of clay and have plastic liners to help prevent leaching. We are lining waste sites containing plastic with more plastic! This was mandated as part of the Clean Air Act through the EPA. Class action lawsuits have been filed as a result of this contamination by those living downstream of older landfills.[140]

If left unchecked, by 2050 there will be 12 billion metric tons of plastic in landfills. That amount is 35,000 times as heavy as the Empire State Building.[141] Effects of plastics on groundwater and the contamination of municipal water systems will be discussed in detail in a future chapter.

We all live downstream from someone and something polluting us. Some of that is poisoning us and the planet we live on. The world produces more than a billion pounds of plastic waste each year and what doesn't end up in landfills ends up in the ocean.

But wait. You recycle. Friends and relatives say they do, too, or at least they try hard. Would you be surprised to learn that 91 percent of plastic isn't recycled? Unfortunately, most of the plastic produced is disposable products ending up as trash.[142]

The sheer quantity of plastic trash produced is horrifying to me. Not only to me but staggering to the scientists who set out to conduct the world's first tally of how much plastic has been produced, discarded, burned, or put in landfills and oceans. "We all knew there was a rapid and extreme increase in plastic production from 1950 until now, but actually quantifying the cumulative number for all plastic ever made was quite shocking," says Jenna Jambeck, a University of Georgia environmental engineer who specializes in studying plastic waste in the oceans. "This kind of increase would 'break' any system that was not prepared for it, and this is why we have seen leakage from global waste systems into the oceans," she says.143

Plastic is also used as packing materials that become trash as soon as the package is received.

What happens to the items put into the recycle bin? Even if you are a conscientious recycler, your plastic may not get recycled. The first stop for your recycled items is a recovery center where glass, plastic, metal, and paper get sorted. Everything gets sorted. After sorting, the plastics are baled and sent to another facility where it is washed, ground, and pelletized. From there, it goes to another facility to be made into a new single use plastic bottle you use and throw away, if it, you, or I follow the rules and are lucky.

It's easy and economical to recycle clean, pure plastic. Unfortunately, more than half of the plastic we put in the recycle bin is "dirty" plastic, plastic that is contaminated by food, paper labels, or other materials. For the past 30 years, the solution to disposing of dirty plastic was to send it to China.

China received ten million tons of plastic per year, most coming from developed countries around the world. The plastics were sent to workshops to be recycled into raw materials to feed the appetite of China–the world's factory. This waste was then exported back to where they came from with a new face such as manufactured clothing or toys.[144] Things changed when the documentary "Plastic China" was made in 2016. The documentary illustrates the truth about contaminated plastic sold to China. It tells the story of a poor Chinese family trying to eke out a living by hand sorting mountains of plastic. China is now waking up to the idea they do not want to be the world's dumping ground for plastic waste.[145]

Unfortunately for the United States, but fortunately for China by some people's thinking, as of January 1, 2018, China stopped accepting other countries' waste plastic unless it is impossibly pure–not more than 0.5 percent of foreign matter. As you might imagine, this is a hopelessly high standard to meet. So, what happens to all the plastic sent to recycling centers not 99.5 percent pure? It piles up in landfills. Most of what doesn't pile up ends in the water.

Now that sending our trash plastic to other countries is less of an option, what happens to all this waste plastic? Ultimately, recyclers have to decide how long they will keep the plastic that cannot be sent away for recycling and how and where to get rid of it. Much of it will go to a landfill. Only large recycling facilities can adapt to the new policy and begin to recycle the "dirty plastic" into plastic pellets. However, this is the exception rather than the rule. Smaller recycling centers are going out of business. It's too costly for them to do business. The choices are to pay to get rid of it in a landfill, figure out how to make pellets out of plastic, or it ends up going out to sea.

Remember, even if you or they think most people recycle plastic, only about 9 percent of all plastic gets recycled.[146] Unfortunately, even if you use the recycle bin, most plastic may never actually be recycled.

Which leads us back to the original problem. Landfill waste, including unrecyclable plastics get dumped into the ocean and ultimately onto our dinner tables and into glasses of water. Let's examine how plastic ends up in the ocean and in your next meal.

A study was launched in 2015, in an attempt to determine the amount of plastic ending up in the seas and the harm it is causing to birds, marine animals, and fish. The prediction made was that by mid-century, the oceans will contain more plastic waste than fish, ton for ton.[147] This has become one of the most often mentioned statements and mandates to do something about. Now, please.

A 2017 study published in the peer-reviewed journal Science Advances, was the first global analysis of all plastics ever made and their fate. Of the 8.3 billion metric tons that has been produced, only 9 percent has been recycled while 6.3 billion metric tons has become plastic waste. The vast majority, almost 80 percent, is accumulating in landfills or is in the environment as litter. What this means is at some point, much of it ends up in the ocean, the final trash dump.[148]

It has been reported that half of all plastic manufactured becomes trash in less than a year. Much of this growth in plastic production is due to the increased use of plastic packaging. Plastic packaging accounts for more than 40 percent of non-fiber plastic.[149] In the 2015 study, it was estimated that 8 million metric tons of plastic ends up in the oceans every year.[150] To put this into perspective, it is the equivalent to five grocery bags of plastic trash for every foot of coastline around the globe. In my opinion, this is due to the rapid acceleration of plastic manufacturing which has doubled roughly every 15 years, outpacing nearly every other man-made material. It is unlike virtually every other material due to its non-biodegradable nature.

All plastic trash ends up somewhere on Earth. A small amount, not enough to make a significant reduction or difference in overall planetary pollution, is burned, fouling the sky. Even a little more smoke is too much pollution in the sky. Somewhat more plastic ends up in the water polluting what we drink and what we are made of. Still, most of what is left becomes landfill, and more ends up in the water. Do you feel more toxic and unwell? Or do you feel more protected and safe?

When plastic enters the environment, it stays there. It stays in the landfills. It stays in the ocean. In a 2018 investigative report by CBS news, it was found that every year somewhere between 5 and 12 million metric tons of plastic waste enters the ocean.[151]

Plastic in the ocean will break down into other smaller pieces. Sunlight can break down some plastics into microscopic bits known as "microplastics."[152] It's these microplastics and the possible consequences of dumping plastics into the ocean we will focus on next.

As mentioned earlier, I am of the view there is only one ocean. We all depend on the ocean for our survival, and the survival of Earth's living organisms and systems. We all need and use the Earth's water. The waters of Earth communicate with each other via currents, not only affecting water flow but affecting our climate.

Think of the ocean currents as giant conveyors flowing through the oceans moving with them huge amounts of water and waste contents constantly. Wind drives surface oceanic currents. Deep ocean currents are mostly influenced by surface temperature and salt gradients.[153]

The North Pacific subtropical gyre is an ocean current created by the interaction of the California, North Equatorial, Kuroshiro, and North Pacific currents. An ocean gyre is a system of circular ocean currents formed by the rotation of the planet and the Earth's wind patterns. An area of intense interest and research is the Great Pacific Garbage Patch.

The Great Pacific Garbage Patch is bounded by the North Pacific Subtropical Gyre. The idea of a "garbage patch" conjures images of an island of trash floating on the ocean fowling the world's water. Whether the trash is piled high enough to block an island view or composed entirely of microplastics, both are associated with many health problems, including cancer. The fish, mammals, birds, and other critters living there get contaminated by toxic garbage. In reality, the "patch" is almost entirely made up of tiny plastic bits called microplastics. Microplastics can't always be seen by the naked eye or even satellite imagery. A recent study of hundreds of plastic drinking water bottles reveal microplastic contamination in the contents of more than nine of ten bottles. Industries making these bottles say there is no evidence of harm from microplastic particles. There is no evidence of microplastic safety in people either. Safe? Harmful? No one knows if long term health improves, or further damage is done to humans by microplastics. The question of microplastic toxicity remains unresolved. More than two billion of the world's seven and one half billion people drink water contaminated with urine, feces, or both. Let us clean the world's water before adding another substance with unknown safety or toxicity. The microplastics of the Great Pacific Garbage Patch may make the water look cloudy like soup. This "soup" is intermixed with larger items, such as fishing gear, shoes, and other plastic items. The Great Pacific Garbage Patch is estimated to be one or two times as big as Texas.[154]

The seafloor beneath the Great Pacific Garbage Patch is also an underwater trash heap. Oceanographers and ecologists recently discovered about 70 percent of marine debris actually sinks to the bottom of the ocean. People make chemical wastes and plastics nature can't breakdown. Whales, porpoises, sea birds, and other sea creatures have been killed by plastic trash. More than fifty percent of ocean pollution begins as plastic land pollution.[155] This fact bears repeating–at this rate, the World Economic Forum predicts that by 2050, our oceans will contain more plastic than fish.[156]

Fish and other edible sea creatures take up microplastics into their systems and enter the food chain. Smaller creatures get eaten by larger creatures; larger creatures get eaten by humans. We know plastic ends up on our dinner plates in our food, in some sea salts, even coming out of your freshwater tap.[157]

When I think of water pollution, I not only worry about the humans on the planet, I worry about other species as well. Being a one-time resident of San Francisco, I am particularly fond of the San Francisco Bay Area. The impact of plastic pollution is clear in the San Francisco Bay Area. The Marine Mammal Center treats 10 to 20 animals annually who are either injured or have ingested ocean garbage. Plastic can also present a problem to whales. A 51-foot whale died from ingesting 450 pounds of ocean trash.[158]

We are now in a situation where we have to play catch up. Planetary societies collectively are far behind the Earth's clean water minimums for both quality issues and simply maintaining sufficient quantities for all needs and uses. There was a lack of awareness of the implications of plastic ending up in our environment until it was already there. Trying to gain control of plastic waste is a large task that will require a comprehensive, global approach. We are going to have to re-evaluate plastic chemistry, product design, recycling strategies, and consumer uses. The United States ranks behind Europe (30 percent) and China (25 percent) in recycling. Recycling in the United States has remained at nine percent since 2012.[159] It will take effort from *every* stakeholder, from governments, citizens, countries, industries, including waste management companies, at every level, to fix this problem.

In my opinion, certain plastics and packaging need to be banned. Even though we can't imagine life without plastic, we won't have life to imagine if we keep living the way we are now. We need to change our disposable culture mentality. Culture is shared consciousness. Does everyone sharing that consciousness wish to abandon that culture? As a society, we need to consider tradeoffs of some conveniences for a clean, healthy environment. We have to think about using different materials for some products that are especially problematic.

I have to admit there are many times I feel we have passed the tipping point for saving our planet's water. Water "futures," buying and selling, has now become a business stock market reality. We can now purchase clean water. How long will that opportunity last? Asking myself that question makes me thirsty! There are many groups trying to address the pollution in the ocean. One holy grail would be to find and implement a solution to plastic water waste.

There was a recent discovery by researchers at the University of Portsmouth in Great Britain. Could a small enzyme be the solution to plastic pollution? The researchers figured out a way for a natural enzyme to digest plastics commonly used in plastic bottles. The enzyme is particularly unique because it digests material that is man-made. Usually it takes decades, if not centuries, for plastics to degrade. The enzyme can break down plastics in a few days.[160]

That's the good news. The bad news is there is still a long development and scaling phase before this small enzyme could help reduce our plastic problem. Currently the enzyme only works on one type of plastic. It is a long time ahead of us before the enzyme could, or if it can, be used on a large scale.[161]

Another interesting discovery from a team of European scientists is a common insect that can chew holes in plastic shopping bags. The insect is the wax worm (*Galleria mellonella)*. In its larval form, wax worms live on the wax in beehives.

Wax is a polymer like plastic, and both have a similar carbon backbone. When left inside a plastic shopping bag, the worms chew holes in the bags. Could they help us rid our plastic bag waste? At the rate the worms chew it would take 100 worms almost 30 days to breakdown an average plastic shopping bag.[162] Interesting, but these possible solutions are still far from solving our plastic woes.

On a much larger scale, an inventive Dutch person is working on a device that will skim the Great Pacific Garbage Patch. Its inventor thinks it will help clean the ocean and hopes to deploy 60 such units.

The device is 2000 feet of plastic piping with a 10-foot nylon screen attached. Plastic again! The skimmer will be towed to the Great Pacific Garbage Patch. If things go as planned, the device is designed to use the wind, waves, and water currents to skim the plastic and corral it into an area where it can be removed. If it is successful, it should be able to remove half this Great Pacific Garbage Patch every five years.[163]

The latest reports on the device show it is not working the way it was designed to work and has not swept up any plastic waste despite being deployed in September 2018. The problem lies in the speed of the solar-powered skimmer. Its slow speed isn't allowing it to hold onto the plastic it catches. Engineers are working to fix the problem, but for now it remains a multimillion-dollar experiment.[164]

Certainly, these are ambitious and lofty goals. My question is – what do we do with the plastic that's skimmed out of the ocean? And so, the continuing dilemma of what to do with the waste plastic must begin anew again.

If you take the time to view the "60 Minutes" segment on the skimming idea, you will hear what I found to be a very insightful analogy of the problem we are facing. It was stated that trying to clean up the ocean with devices, is like trying to mop up an overflowing bathtub with the water running rather than trying to turn off the tap.[165]

This leads me to thoughts of "turning off the tap." What do I reach for out of convenience rather than necessity? I try to make healthy choices of what I put into and onto my body. Especially what I drink. I carefully evaluate what I am willing to abandon to reduce my dependence on plastic that quickly becomes trash ultimately put into the water and the rest of the environment.

Until the enzyme, the wax worms, the skimmers, and more effective interventions can be deployed to help reduce the plastic pollution of the ocean, what can we all do in the meantime? Here are some suggestions:

Reduce or eliminate your use of single-use plastics. These items include plastic bags, soda bottles, water bottles, straws, cups, utensils, dry cleaning bags, takeout containers, plastic tableware, and all plastic items that are used once and thrown away. Thrown away means most plastic ends up in water somewhere on Earth. Refuse any single-use items you don't need (straws, bags, etc.). Purchase and take your reusable types of these items. Let the businesses you frequent know you would like them to offer options other than single use plastic items.[166]

Recycle properly. We've already examined the problems with recycling certain plastics. Do your part in properly recycling when you can. Be sure to check with your recycling center about the types of plastics they accept.[167]

Consider participating in a beach or local waterway cleanup. This helps reduce plastics in the ocean by preventing them from initially getting into the ocean. This is one of the most direct ways to fight plastic pollution. I encourage you to look for local events in your area. It can be very rewarding. But again, where do we put the items we remove?[168]

Support bans on single use plastic bags, takeout containers, and bottles. Carry your own reusable grocery bags. It may seem like a hassle, but it makes a difference![169]

Avoid products containing tiny plastic particles, called "microbeads." These small plastic particles have become a growing source of ocean plastic pollution over recent years. Microbeads can be found in some facial scrubs, bodywashes, and toothpastes. They readily enter the waterways through sewer systems. They have the potential to affect hundreds of marine species. Look for ingredients like "polyethelene" and "polypropylene" on the labels and avoid these products.[170]

Become an advocate for the ocean. By reading this book, I know you are interested in water. I hope you are becoming motivated to make changes having personal, local, and global impact. Continue to stay informed on issues related to plastic pollution and help to make others aware of the problem. Encourage your friends and family about how they can be part of the solution. Consider watching plastic pollution focused documentaries like *Bag It,*[171] *Addicted to Plastic,*[172] or *Garbage Island.*[173]

Consider supporting organizations addressing plastic pollution. Many non-profit organizations are working to reduce and eliminate plastic pollution from our ocean. Here are a few to research - Oceanic Society, Plastic Pollution Coalition, 5 Gyres, and the Plastic Soup Foundation to name only a few.[174]

Water makes and does various things to help and make life possible. We must work together. Stop dirtying water. The Earth would solve her own water issues if people leave her alone. Earth has fixed water problems before (e.g., earthquakes, floods, ice ages, storms, and more). Through water, Earth will achieve success again. She might have to do it after humans pass the tipping point or our species is long gone. Please help prove me wrong. I don't want to be right about pollution and weather calamity in the water and the world. Let's work together to clean the water. The more we try to do both, the more neither becomes either just when we need benefits from each. Stopping the continued dirtying of water is a huge job for the world's population. Everyone is involved. Cleaning the world's water is an enormous task. We are getting farther from that goal. Each job must be done. There is more water than people, there always has been. I think there is still time for desperate frenzied behavior. That time is getting smaller.

Looking at the big picture, we're one people, one ocean, one Earth. It will take every current participant's best effort to fix the world's water that has been left full of collective plastic and pollution problems. Every person and every life form currently alive is a colleague in cause. Worst case scenario, our children and their offspring may miss their chances to fight in the last war for their cultures water. The water problems cannot be fixed here or anywhere until after we identify and execute unknown future plastic solutions and fix the water pollution problems. We need to think farther upstream and further into the future to help find solutions and stop causing more problems. The solutions would not happen fast enough.

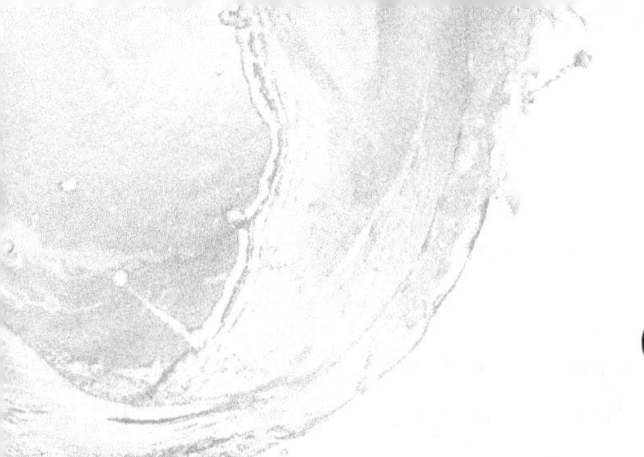

CHAPTER 7

Having no water is bad. Having no water is worse than having bad water. Even drinking bad water is better than having no water. Neither is a good choice to have.

Fishing in the world's oceans and waters not only devastates many fish populations but adds more ghost gear to the long-lasting plastic waste. From the smallest kid stream fishing to the largest fleet of fishing vessels scanning the oceans, each adds to the plastic waste in water. Everything from fishing line, hooks, sinkers, bait, nets, empty water bottles, and more gets left in the water. Snags, rocks, sunken ships, and other underwater obstacles exact their toll on fishing supplies. Lost phantom materials swallowed by sea life can end the lives of whales and many other animals, both big and small.

Previously, we explored the problems of dumping our junk into the ocean. To address water problems, start where the problem begins, upstream figuratively and literally. In my opinion, a glass of public tap water from most anywhere tells the true story of oceans, lakes, rivers, streams, aquifers, water supplies, and rain drops there and everywhere.

Could that water have been clean and pure? How long ago was that? How far into the future will it be again?

Providing clean water can be done. Some cities in developed countries, big and small, have done it. What if we could bring all the world's water, at least up to minimum standard of safety, cleanliness, and purity? Those hopeless water standards must be developed first.

I believe there's no clean, uncontaminated freshwater or saltwater anywhere in the world. Why do I think this way? We live on a planet where everything is interconnected - air, water, land. We know the Polar ice caps get fouled by air pollution.[175] Industries and individuals everywhere add toxins to the saturated mix of water pollutants. The parts of the Earth not covered with water are kept alive with water. This is especially problematic if you believe as I do that all water in the environment is polluted to some degree. Water goes everywhere. Water escorts pollution traveling with no boundaries too.

As mentioned, I grew up in the southern part of New Jersey on the riverside. The Delaware Water Gap and Delaware River are to the west, flowing south forming the western New Jersey boundary. They flow into the Delaware Bay and Atlantic Ocean. The Hudson and East Rivers aren't in New Jersey, but they're to its north then east flowing south finally into the Atlantic Ocean. These waterways help define the state's land borders, outline the water geography, and provide some drinking water supply for much of New England. After climbing north through the rural upstate land of New York, is the water safer at the river's headwaters? Or where it's currently leaving the east coast to join the Atlantic Ocean? I mean ocean only. Get local information. You decide.

In my mind, rivers mirror the flow of value and threat (e.g., valuable essential water and harmful poisonous deadly toxins into our body water, and lives). Between the time I grew up and now, rivers seem to be a shadow of their former state if you look at the clarity, color, content, birds, and marine life. When viewed from above, the waters look like dirty roads with trash along their sides. There are papers, bottles, and miscellaneous debris along what should be the empty clean shoreline. That pollution can be seen and the water contains more toxins that cannot be seen.

Polluters dump unknown chemicals into the Garden State. Should people drink water or eat food from that "garden"? Millions of people do both. I fear that most of those millions are unaware of the potential pollutants they consume or what health problems these hazardous wastes will cause.

The central and coastal portions of southern New Jersey contain the Pine Barrens. Running beneath this pristine forest is part of the huge Kirkwood-Cohansey aquifer. It's the largest ground water system in New Jersey. Central New Jersey and the Pine Barrens area is a huge natural water filtration system. The soil and sand of the aquifer remove toxins as the water flows up from under the ground. This is where the term "groundwater" comes from. All water is a valuable resource of which we should be aware. Despite the Delaware River to the west and the Atlantic Ocean to the east, all rivers and streams originate in and flow out of the Pine Barrens, none flow in from the Delaware River or Atlantic Ocean.

This and other water systems are threatened by a variety of toxic chemicals. Many different chemicals are used in various manufacturing processes in general. Chemicals are used in hydraulic fracturing, better known as fracking. Fracking is not just a New Jersey Issue.[176] This matter has both benefits and inherent risks. Fracking occurs in 34 states in the US, according to a 2016 analysis.[177] As the dangers of fracking become known, public anti-fracking pressures have increased. Three states–Vermont, New York, and Maryland–have banned fracking.[178]

Fracking is a process used to extract gas and oil from shale rock. It is the process of drilling into the earth and pumping under high-pressure, a mixture of water, sand, and chemicals in order to fracture the shale rock and release the gas inside. The drilling can be vertical or horizontal. The name "fracking" comes from the fact that the rock is fractured apart by the high-pressure water and chemical mixture.[179]

Many chemicals are used in the fracking process. Additionally, a lot of water is used too. First, let's look at the chemicals being introduced into the soil in this process. Finding out what chemicals are used is difficult. Petroleum companies consider their fracking mixtures to be proprietary information or "trade secrets." Here are some issues to consider. Gas and oil companies can choose from over 600 chemicals to use in their fracking mixture. Typically, they use about 5 to 10 chemicals allowed by local law in the process. That may not sound too bad - over 600 reduced to 10, but the problem lies in the number of wells that are fracked. Multiply 10 or so by ten thousand and therein lies the potential for problems happening to water, including drinking water, to our health, and to the environment.[180]

Chemicals used in fracking include acids, like hydrochloric acid, that dissolve minerals in the rock. Agents like quaternary ammonium chloride and glutaraldehyde are used to prevent corrosion and eliminate corrosive bacteria to protect the drilling equipment. Other chemicals aid in hydrocarbon extraction, winterize the fracking fluid, and act as a non-emulsifier. These chemicals can include surfactants like ethanol, methanol, lauryl sulfate, and others.[181]

Additionally, analysis of fracking fluid has identified the presence of volatile organic compounds (VOCs) such as benzene, toluene, ethylbenzene, and xylene–all of which are known human carcinogens. Carcinogens are cancer causing agents.[182] More on the dangers of benzene later in this chapter.

Although the percentage may be small, the number and types of chemicals play a crucial role in fracking. For the petroleum industry's perspective on the use and safety of the process and chemicals used in fracking, I direct you to the website FracFocus.[183] The companies using these chemicals and fouling the water provide this information to assure the public that fracking is safe. I remain skeptical.

As already mentioned, the main component of the fracking mixture is water. A lot of water. Now, let's look at the water portion of this process.

An often-used action in fracking is the use of slick water to aid in the fracturing and recovery process. Basically, this method involves water combined with a polyacrylamide friction reducer. The slickening agents reduce the friction of the water in the pipe and the thickness or viscosity of the water mixture. When water is less viscous, it is lighter, and more volume is needed to carry the same amount of sand to keep the fracture network open. Because there is more water to move, higher pressures are needed to transport it. This higher pressure helps to create more fractures which allows the natural gas to flow back up the well more quickly.[184]

There are many worries about fracking. One of the biggest being the volume of water required per well. Different sources give differing amounts of water required per well, Remember, there are tens of thousands of wells operated by each of the many petroleum producing companies. Sources are quick to say that there is no "typical" well water volume but estimates for water per well range from 1.5 to 16 million gallons of water.[185] As water quality and conservation becomes a larger issue, fracking takes on even more significance. Adding to the problem is that in many of the areas where fracking is taking place (e.g., Texas and Oklahoma, which constitute over half of the nation's wells), there is serious drought. There is concern in these areas about fresh water taken from community reservoirs and environmental ground waters.[186]

You may be asking yourself what happens to all that water? Is it lost? Is it returned to the water supply providing your drinking water? Should you take the chance repeatedly, day after day, and drink that water? Should adults bathe babies, children, or themselves in this water even if it's somehow "treated"? What if treatment has changed and the current treatment regimen is ineffective, unsafe, or damaging? What if some chemical treatments may be toxic or harmful? As much as 60 percent of the water injected into the well during the fracking process will be recovered as wastewater requiring either recycling or disposal. For the life of the well it is estimated that approximately 100,000 gallons of wastewater per day will be produced.[187] This water can create concerns with the local drinking and agricultural water supply.

Petroleum producers are facing more stringent regulation about the recycling and disposal of wastewater from fracking wells. More on the regulations and Clean Water Act will come later. Regardless of regulations, it is my opinion that spending money may influence officials to buy commercials on television and elsewhere to decrease consumer concerns about water availability and uses, but the problem remains; what is done with the wastewater?

Trying to deal with the vast amount of water being used and needing to be recycled is creating a booming industry. Companies need to source fresh water, transfer it to a frack site, then dispose of the end result. Solutions to produce fresh water such as reverse osmosis, evaporation, electro-coagulation, and chemical coagulation are being explored.

The quality of water recycled varies significantly in different parts of the United States and the world. Many experts believe technologies should not be selected solely by operating costs, but also by efficacy to certain types of water, or water processing. Still, I remain suspicious and unconvinced about the effectiveness of various interventions. No matter how big and important, or small and seemingly insignificant, act and do what works for your local situation.

One method used to recycle the wastewater is on-site oilfield recycling. Benefits cited with the use of this method include a relatively low operating cost for processing of chemical coagulation, along with a process called dissolved air floatation. The chemical coagulation and dissolved air floatation system takes the contaminated fracking wastewater, eliminates most of the suspended solids (remember, sand is used in the fracking process), and kills about 99 percent of bacteria. This method allows well operators to use recycled water instead of fresh water from local water supplies. It has been reported this method can offer significant savings to the petroleum companies, ranging from

$100,000 to $500,000, when taking all factors of water acquisition and disposal into consideration.[188]

A good water recycling system may provide a realistic solution for water conservation in the petroleum industry. Each system can recycle upwards of 20,000 to 30,000 barrels of frack wastewater per day. This can translate into a savings of approximately $1 to $2 per barrel for the well operator. If a well uses 5 million gallons of water, the savings add up quickly.[189]

It is beyond the scope of this book to explore all of the consequences of fracking (e.g., earthquakes, water contamination, release of methane gas, etc.). I encourage you to read more on this subject if you are interested in learning more about hydraulic fracking.

Bad news about water in New Jersey never seems to end. I can recall a train derailment in the winter of 2012. News reports showed train cars derailed from a bridge in southern New Jersey. It was said the train engineer went through a red stop signal and crossed an unstable bridge. Some cars following the engine fell from the bridge and crashed into the water of Mantua creek near Paulsboro. Television news showed tank cars in the creek. Emptying from the split tankers was the chemical vinyl chloride. Each of the railroad cars carried thousands of gallons of vinyl chloride. It was estimated approximately 23,000 gallons of vinyl chloride dumped into the creek.[190]

All users need to care more for water.

Vinyl chloride is a man-made, colorless, toxic gas used in the production of polyvinyl chloride (PVC). PVC is used to make a variety of plastics. Minor exposure causes nausea, vomiting, and dizziness. Major contact can produce blood in the lungs, cancer, and death.[191] Exposure to vinyl chloride is associated with an increased risk of blood cancers, as well as brain, lung, and a rare form of liver cancer.[192] Definitions of minor and major exposure vary among health care practitioners. Some people, like me, think any exposure to a chemical causing cancerous growth is too much risk for even small single exposures. Other people think clean uncontaminated water minimizes that hazard. Minimal exposures can be associated with dangerous harmful effects. A cancer-causing chemical is being used to make more things that may be associated with cancer. That chemical will go everywhere water and air goes. Does every spill anywhere over a long enough time endanger some people on Earth somewhere? I think it does.

A mandatory evacuation of the area was ordered. Vinyl chloride contaminated the water, and its vapors contaminated the air. It polluted the town and the people. Safer, double hulled tank cars are available and are mandated for all new-build oil carrying tank cars. Vinyl chloride and other hazardous chemicals are required to be transported in a double hulled tank car.[193] Vinyl chloride is just one example of a toxic chemical transported by rail all across the country. Many other dangerous chemicals are shipped along America's highways and railways and are spilled as derailments continue to occur. The public grows more aware and alarmed at what is being transported near their homes.

Water from Mantua creek empties into the Delaware River, then the Delaware Bay, and finally into the Atlantic Ocean. In my opinion, nothing can be "contained" after it contaminates running water or flowing air. All the vinyl chloride wasn't recovered from the train spill. It couldn't be. Vinyl chloride is a gas that began dissipating as soon as it was released from the car into the creek following into the atmosphere. The surrounding local area was under an evacuation order while the Coast Guard and other authorities tried to contain and remove the vinyl chloride. Unfortunately, cleaning up vinyl chloride presents a logistical challenge. Using acetone and suctioning out the vapors from the breeched tank cars, the Coast Guard attempted to remove the toxic chemical.[194] When authorities "clean" the spill, their cleaning can end up creating a new problem by using another known toxin, acetone. Like the debris retrieved from the ocean, how is the cleanup material disposed of? Does it stay "contained"? Where is it contained? Is it no longer toxic? Is it more toxic? What long term damages will occur? How long will the effects last? If deaths happen, will they be linked to the chemical spill or the water? Should they be? What about longer-term cancer deaths? Let's find out and be sure. Make water safer.

I met Meghan when she was a newborn child of my friends. Years later, Meghan, her parents, and I were walking along the deserted part of the southern New Jersey Delaware River shoreline. While hiking, a tank ship passed us heading north up-river to Philadelphia. At the time, no one knew it was leaking. "What's that smell?" Meghan asked. The odor got stronger as we got closer to the large spill that no one could see. Her parents and I recognized the smell of benzene. We left the beach and started the walk home.

From that spill, benzene entered the water table and surrounding land. Benzene flowed into drinking water sources. Only tiny amounts, but people unknowingly drank it, bathed in it, swam in it, and played in it, long after the odor escaped the water. Evidence of the benzene spill is unable to be found, or difficult to find in newspaper records or on the Internet. It vanished with other noxious, deadly, and unprofitable news stories. I was there when it happened. I saw the effects. Thousands of dead fish were scattered on the riverbanks. Stories of dead fish I saw never made the news.

I believe Meghan's untimely death began with toxic benzene spilling into and spreading through river water, then exploding through her bone marrow and causing cancer. The river accident was never connected to Meghan's three cancers or her death. Meghan died from cancer ten years after her benzene exposure from water at age six. Meghan got three different cancers during her short sixteen-year life. Was benzene exposure the cause of the cancer that ended her life? Will we ever know for sure? Have other people been exposed to benzene without any adverse effects or cancers produced?

I believe there's no safe amount of any cancer-causing substance in water or people. A small toxin amount may cause cancer in some people while a larger amount doesn't always cause cancer in others. Everyone knows an eighty-year-old who has smoked two packs of cigarettes a day for sixty years and doesn't have cancer or heart disease. Everyone knows a twenty-year-old who has smoked since age ten, gets cancer, and dies. Neither of these examples is uncommon. We know if you expose yourself to tobacco, you are increasing your risk, and no one knows exactly how much or over what time-period. The same is true for the risks associated with toxic water. Every person is a group of genes different from every human being who has ever, or will ever, live. As if being born with different genes wasn't enough, continual epigenetic changes occur driving more changes in chromosomal expression. No matter how much people are the same, every person is different from everyone else. Beneath those differences breathes genetic similarity but not the same identity. No two twins are the same. Time changes everyone.

One person's chance of getting or dying from a disease is different from every other person's chance. Chances are that your risk is somewhere in between. Risk is individualized. Your risk is your risk and no one else's. I believe factors playing a part in your risk come from the water you drink, your host defense mechanisms, and current functional competences, which add up to make your good health.

Every bit of water has potential to accumulate poisons. Bad water is everywhere, from the ocean to the smallest amount of rain, snow, puddle on the ground, clouds, or fog in the sky. This problem can be caused by various chemicals, radiation, bacteria, viruses, and toxins in water, air, and food. Prolonged exposure to specific and wide-ranging factors can cause cancer to occur. It seems too much of most everything at the wrong time can cause cancer in an "at risk" person. Risk comes from many sources. The precise amount of each chemical in water to determine if it's safe or toxic to certain individuals is not known. Books are available to help you educate yourself.

Water tragedies may be happening where you live. Are you affected? Whether affected or not, be angry. Be aware. Stay as safe as you can.

Unfortunately, situations such as these are not limited to New Jersey. Examples can be found all over the country and throughout the world. They are occurring more out of control and doing more damage.

Another example touching my life occurred outside of Philadelphia, where I did my medical training and internship. A large chemical spill happened in the Schuylkill River. The Schuylkill River provides drinking water for people in Southeastern Pennsylvania. It empties into the Delaware River. Both rivers provide drinking water for millions of people.[195]

Years ago, I can recall tens of thousands of gallons of deadly PCBs were spilled into the Schuylkill River caused by a crash of two oil tankers in the Delaware River.[196] I was concerned by the spill of this toxic substance in an area supplying drinking water to millions of people.

These accidents are not the only examples of severe water contamination. Some of you may remember when the Cuyahoga River in Cleveland was burning. In June of 1969, an oil slick, due to years of pollution, caught fire on the Cuyahoga River. The fire in 1969 was not the first time the river burned. This time, though, the fire ended up on the cover of Time Magazine and in other news outlets. Bright red flames rising more than a hundred feet into the sky made for an impressive sight, though no pictures of the 1969 fire are known to exist.[197]

How can a river made from water be on fire? How many drinking water sources does the Cuyahoga River supply? Do the toxins flow into drinking water sources downstream? Is someone drinking them? Is a child bathing in them? Are pregnant women drinking contaminated water? Are hospitals using those water sources to help get sick people better?

The memory of seeing a river on fire leaves me with a mental picture making me lose my position in time and space. A river on fire makes me not know where I am or why I am there. The picture is in my mind whether my eyes are open or closed. The image never fades or goes away. My mind tells me if water can catch fire, nothing is safe. What does a river on fire say to you? Perhaps the chemicals in the water may end more lives than the fire. We can retreat from the fire, but we can't escape chemicals in water and on the land.

We see the effects of the upstream events downstream. Many of these episodes are out of our control (chemical spills, industrial wastes, fracking, strip mining, etc.). It may make us want to throw our hands up and ask what can we do as individuals to help "turn off the toxin tap"? We've discussed the mantra of reduce, reuse, and recycle. Here are a few more ideas to consider along those lines.

I could go on and on – the Chesapeake Bay, Love Canal, and many more reports than can be recounted here. I hope you see we all live downstream from someone or something adversely affecting our water supply. What effect does that have on our health?

When there were fewer than one billion humans on Earth, the worldly expanse was difficult to pollute. There was room for expansion and dilution. Now there are almost eight billion people. Pollution overflows and chokes the entire world with toxins we touch, drink, breathe, and eat. The following are things commonly used, happening, or are purposefully done, causing water pollution and potential harm to people.

1. Soap

Soaps are surfactants, surface active agents used to make wastes dissolve and remove them from anywhere they are. When they're removed, they frequently end up somewhere in a worldly subsection made mostly of water. If you wouldn't drink water with soap in it, I suggest you limit the soap you put into any water you'll eventually drink. Some life form somewhere will drink all the fresh water you see.

2. Automobiles

Gasoline, diesel fuel, windshield washing solution, oil, antifreeze, transmission fluids, brake fluids, and other substances can contribute to water pollution. Those substances leak onto the Earth and find their way into the water table. Properly handle, dispose of, and recycle these chemicals before they become pollutants. All these substances can be found in the water.

3. Pets

The United States alone has millions of dogs and cats. Not all have homes. Animals eliminate waste every day. Pet waste can enter the water supply if not disposed of properly.

4. Livestock

Livestock waste can pollute water that enters the food or water supply. The meat industry raises and kills tens of billions of animals a year for food. Bacteria isolated and transmitted from animals have been shown to be the cause of many food-borne illnesses. We can all recall a news story about an E. Coli, Salmonella, Listeria, or another type of bacterium outbreak that has been linked to a food source either animal, vegetable, fruit, or water. Yes, all types of food and water have sometimes been contaminated causing epidemics, and/or infections.

5. Water Runoff

Water from drains, ditches, and storms can carry toxins into rivers and streams. Whatever is on the property where you live could pollute something and/or someone downstream. Everything running into water could eventually become a contaminant in drinking water. Even many distant deserts get a little rain now and then. Most chemicals put into water stay in the water. Relatively few chemicals are ever totally removed.

6. Septic System

If you live in a rural area you may have a septic system. Household waste flows to the septic tank where bacteria breakdown the biosolids before the runoff, or effluent, enters the drain field, then possibly enters the water supply. Some people believe possibly is eventually given enough storms or just enough time. The effluent still contains potentially disease-causing bacteria. To be safe from contaminating drinking water supplies, a drain field must be 50 to 400 feet away from the nearest water supply, but that depends on the soil and the number of people served by an aquifer.[198] Is this "margin of safety" always abided by? Do we trust everyone to maintain a healthy septic system? Do you want to consume human waste from the water table?

7. Storms and Natural Occurrences

Storms and natural disasters can wreak havoc on the waterways globally and will be the focus of a future chapter. Here are a few examples of how they affect water upstream, downstream, and globally.

Hurricane Sandy, the strongest hurricane in the 2012 American hurricane system caused over 10 billion gallons of either raw or partially treated sewage to be spilled into the waterways of New Jersey and New York. It covered streets and flowed into homes. It was estimated that more than 90 percent of the sewage flowed into the rivers, canals, and bays of these two states. Dunes along the beaches were breached, even subways were flooded.[199] Hurricane Sandy exposed serious shortcomings in our country's aging infrastructure.

Every storm event has the potential to worsen drinking water.

8. Lawns

Water toxins used for lawn care play a role in ecosystem damage. There're always new pesticides, herbicides, and fertilizers, in addition to unknown mixtures of old killers. These chemicals can end up in the water table.

9. Water Creatures

Fish, reptiles, amphibians, waterfowl may die from toxins used on lawns rain washes into streams, or from contact with plastic bottle caps, fishing lines, and other plastic waste they come in contact within the waterways.

10. Boats Large and Small

Numerous pleasure boats and watercraft populate America's lakes, rivers, and bays. These have the potential to pollute water upstream and downstream through various "normal" activities. Responsible boaters should dispose of their trash and sewage properly. Remember, the "fresh" water we swim in contains the sewage and waste of other waterway users. People for both pleasure and commercial purposes use water every day. Water never rests.

I recall an instance where an East coast beach was closed due to the presence of medical waste, including surgical garbage, used needles, and syringes, as a result of dumping garbage offshore. These things have been found on New Jersey and New York beaches. Up until 1992, it was legal for barge loads of waste to be dumped into the Atlantic Ocean off east coast beaches of the United States. Barges loaded with tons of garbage headed into the Atlantic Ocean to dispose of New York City's waste.[200]

If everyone followed the rules, would we still experience legal pollution of our waterways from boaters, both large and small?

11. Household

There are few controls on some menacing chemicals bought and used by individuals. Some chemicals have limits on outside use but are unrestricted for household purchase and use. The size of the household container may be smaller, but you can buy as many containers as you have money for. These chemicals are often used in the household without regard to toxicity. We already know plastic has found its way onto our dinner tables. Have these chemicals found their way there too?

Workers you employ around the house may use a variety of chemicals with unknown short- and long-term toxic reactions. Workers are at your house to get a job done. Workers aren't there to keep you, your children, or pets safe. Often that job is completed without you seeing how any chemicals are used or disposed of where you live.

Often, you don't see what was dumped on your property's watershed area. The toxicity can occur after workers have gone. These substances can get into groundwater. Perhaps you have had a pet sickened or die after a lawn treatment. Many factors could have caused or contributed to the problems discovered, including the common catch all term "unknown causes." As you can see from reading the list, there are things in and out of your control. For those things in our control, try to control them. Reduce use, reuse whatever you can, and recycle responsibly. For those things out of our control, we are all caught "downstream" so to speak. In these cases, be aware of your surroundings and how elected officials, and neighboring industries, can affect our fragile environment. Clean water is Earth's most important sole everyday goal.

If you need motivation, I suggest hiking to as unspoiled a place you can find and let the cleanest water motivate you to be a good local, regional, and global citizen. For me it was a trip to Victoria Falls in Zimbabwe. The falls looked like the center of world and had the purest, cleanest water I had ever seen. I hope it always stays that way.

I would like to see us as a country stop polluting my first home state, New Jersey. While we are acting to stop pollution, do the same for all other states in the US, countries of the whole planet, and water everywhere. There's no other place to go. There is no planet B, C, or D. Chances are, there will be no other planet to escape to, or means to escape no matter your age.

I hope my words connect you to a watery part of the planet you don't get to very frequently. Please stay for a while. Make water more useful to most of Earth' people and animals. Focus on water longer than you took reading this book. Many thanks.

The Antarctic icecap contains 70 percent of the world's fresh water frozen as ice. That ice melts into undrinkable saltwater.

CHAPTER 8

What comes out through your water taps? We've looked at the problem globally. What is happening in the ocean? We've looked at the problem upstream in local waterways. How much do we know about what is coming out of the home faucets? Do mistakes in water systems ever happen? How quickly are they fixed? How soon are we told? How fast were our children warned and protected? How can we protect ourselves from what surrounds us?

In chapter 4, we examined water supplied to our homes by well, private, shared, or by a municipal water supply. Let's look at these options more closely and look at water in, or not part of, our drinking water.

Well water comes from ground water. It is the responsibility of the well owner to ensure the water is safe. Unlike water supplied from a public source, well water does not go through any treatment prior to use. Private wells can be contaminated from various sources both natural and synthetic.

Contaminants commonly found in well water include microorganisms, nitrates and nitrites, heavy metals, organic chemicals, radio-nucleotides, and fluoride among other dangerous substances. Each of these carries risk to human health.[201]

Microorganisms can include bacteria, fungi, parasites, viruses, prions, and other things, some discovered, and many others suspected, but as yet undiscovered. Some not thought of or imagined. Possible sources for these contaminants can be human sewage, animal waste, or expected microorganism growth and evolution. When contaminated water is consumed, people may experience gastrointestinal illnesses and/or infections, or other symptoms and signs. Wells can become contaminated through rainfall or snowmelt runoff carrying microorganisms into the well system or by seeping underground. Effluent from septic systems or leakage of waste from underground storage tanks can reach groundwater sources and can result in microorganisms being present in well water.[202]

Human sewage, animal waste, and chemical fertilizers can cause nitrate and nitrite to enter the groundwater supply. Both nitrate and nitrite are naturally occurring inorganic ions present in the environment. When decomposition of organic material occurs, ammonia is released into the soil. Nitrate and nitrite are produced when ammonia oxidizes or mixes with oxygen. Nitrate is more commonly found than nitrite in soil and in groundwater. Nitrate can also be found in a variety of vegetables like broccoli and cauliflower. In the body, nitrate is converted to nitrite. Drinking water contains nitrates too. If a well has high concentrations of nitrates, a significant exposure to nitrates can occur. High nitrate concentrations can be of particular concern in households with infants who consume formula made from well water or households with pregnant women because of the affect nitrite exposure has on blood's hemoglobin. These substances can reduce the ability of blood to carry oxygen throughout the body. Consuming high concentrations of nitrate/nitrite in drinking water can cause a condition known as methemoglobinemia or "blue baby syndrome" in newborns and infants. Infants below six months of age, drinking water with high levels of nitrate can become seriously ill and die.[203,204]

Heavy metals leach into drinking water from household plumbing and service delivery lines, mining operations, petroleum refineries, electronics manufacturers, municipal waste disposal sites, cement plants, and natural mineral deposits. Heavy metals have numerous sources, both known and unappreciated. Heavy metals include arsenic, antimony, cadmium, chromium, copper, lead, mercury, selenium, and many more. Heavy metals can contaminate private wells through groundwater movement and surface water seepage and run-off. People consuming high amounts of heavy metals risk acute and chronic multiple toxicities, liver, kidney, and intestinal damage, anemia, and cancer. Ultimate risk has much to do with pre-existing underlying health or disease conditions, be they acute or chronic.

Mercury is the universal pollutant. Toxic mercury pollutes all the Earth's water. Water is the universal solvent. Water helped make the Grand Canyon.

The world puts more than eighty thousand pounds of mercury into the air every year. Knowing that should make everyone short of breath. Mercury fouls the atmosphere from burning coal to produce electricity.[205] That mercury continually falls from the air into all the world's water. After the Fukushima nuclear accident, Japan burned more coal and other fossil fuels to make electricity. This added more mercury and other toxic metals to the world's air supply everyone breathes.

China, Canada, the United States, and other countries say they're getting the message about mercury contamination in air and water and are reducing coal burning to generate electricity. Don't believe them until mercury problems in the world's humans start to decrease. Mercury problems in people are still increasing.

One out of every four people surveyed in America thinks "silver" fillings in their teeth contain mostly silver. "Silver" dental fillings contain mostly (49.6 percent) toxic mercury. Mercury amalgam has been the dental filling of choice in the United States for more than a century. It is still.

In Europe, during July of 2018, the use of mercury amalgam tooth fillings was prohibited in children under fifteen years old, and pregnant or nursing women.

American dental practices create many tons of mercury waste each year. Mercury containing dental amalgams are the biggest cause of mercury water pollution in the country. Stopping the use of dental amalgams containing mercury will reduce mercury toxicity in people and mercury concentrations in public drinking water.

The FDA allows dentists to put mercury containing fillings into a pregnant woman's or child's mouth. Mercury in pregnant women, in all women, and mercury in kids and other people, is always harmful. Insist on guidelines motivating people to do the right thing for the water and all people everywhere using mercury. The FDA's policies about the accumulation of mercury help most people but aren't consistent and may lead to more poisoning of some people living in the United States.

Dental staff, including dentists and technicians, wear protective masks, gowns, gloves, and shoe covers. Patients don't. Should they? The dental staff leaves the room while shooting x-rays through your mouth and into your unshielded and unprotected brain. Has either behavior been proven safe for us or them?

Take an active role in deciding what is put into your mouth, your children's mouth, and the water supply. Decide if, when, how, and how much mercury needs to come out of your mouth. All mercury is toxic to people.

The FDA warns pregnant females not to eat excess amounts of high mercury containing fish.[206] Tuna, tilefish, mackerel, sharks, swordfish, and others contain elevated concentrations of mercury. These aren't the only fish high in mercury. They are among the worst. All freshwater fish contain mercury. The larger the fish size, the more mercury it contains.

During the 1950's and 1960's a new disease was discovered in Minamata, Japan. The Chisso corporation had dumped twenty-seven tons of mercury into the water of Minamata Bay. The mercury contaminated the sea creatures. Local residents ate the fish and shellfish. Eating mercury-contaminated seafood causes Minamata disease. Minamata disease is mercury toxicity. Signs and symptoms of the disease may include ataxia, numbness in the hands and feet, general muscle weakness, loss of peripheral vision, and damage to hearing and speech. In extreme cases, insanity, paralysis, and/or coma may occur, and death follows within weeks of the onset of symptoms. A congenital form of the disease can also affect fetuses in the womb. Three to ten thousand people were sickened by Minamata disease. Thousands died. Cases continue to develop and are found more than sixty-five years later. To date, there is no known cure.[207] Stop eating the fish!

All fish accumulate mercury. The bigger the fish, the more mercury it contains. Eating those fish is also bad for non-pregnant people. The non-pregnant people haven't been effectively warned about mercury and water. It's estimated ninety percent of the world's large fish are gone. Worldwide, there's more illegal fishing than legal fishing. Pollution and over-fishing are killing fish and other water life forms. Some fishing methods make the water more toxic.

All water contains organic chemicals, both helpful and harmful. They are found in many house-hold products and are used widely for agriculture and industrial uses. They can be found everywhere, in parks, dyes, pesticides, paints, pharmaceuticals, solvents, petroleum products, sealants, and disinfectants. Organic chemicals can enter ground water and contaminate private wells through waste disposal sites, spills, and surface water run-off.[208] People consuming high amounts of organic chemicals may suffer from damage to their kidneys, liver, circulatory system, nervous system, and reproductive system.

Radio nucleotides are nuclear forms of elements such as uranium and radium. They are harmful to humans and can be released into the environment from uranium mining and milling, coal mining, and nuclear power production methods. Radio nucleotides may also be naturally present in ground water in some areas. Radio nucleotides can contaminate private wells through groundwater flow, wastewater seepage, and flooding. Drinking water with radio nucleotides can cause toxic kidney effects and increase the risk of cancer.

Before the 2012 train wreck in New Jersey, another environmental damaging event happened farther from home. On March 11, 2011, an earthquake measuring 9.0 on the Richter scale struck Japan.[209] A tsunami wave measuring thirty-three feet high further damaged the horrible situation. The wave swamped the Dai Ichi nuclear power plant in Fukushima.

A meltdown of three radioactive nuclear power cores followed. The reactor cores were burning through the Earth. Hundreds of tons or radioactive water containing more nucleotides still enter the Pacific Ocean every day. The effects of radiation in Japan, on its local people, on the world's water, and on the rest of the planet are theorized, but not known. The extent and effects from the damage are not over.

The radiation effects from nuclear bombs and nuclear "accidents," including Fukushima, Chernobyl, Hiroshima, Nagasaki, Three Mile Island, Santa Susana, and others last forever. Forever is at least tens of thousands of years.

If radiation from these accidents lasts that long, people at all levels should ensure that accidents do not happen or can be contained. You and I will not be alive then. Will any humans be alive then? Radiation from Fukushima has been found on West Coast United States beaches. Some water damages on Earth cause immediate death, while others can cause death over time.

Early warnings about dangers from drinking fluoridated water were provided in the 1950's. The toxic fatal reactions from consuming excess fluoride, including cancers in animals and people, have been known since then. The amount of fluoride in your body depends on the amount of fluoridated water and products you drink. No one is routinely monitoring toxic fluoride amounts in all children or adults. To support good health, don't expose yourself to excess fluoride.

Most people in the United States use fluoridated water. Some people may not realize that. Fluoride is added to two of every three United States public drinking water systems. Most countries in Europe (about 98 percent) reject water fluoridation. Dental health in each area is the same. Opposite fluoride actions in one study by different countries produce the same outcomes in people's teeth. Why don't fluoride using Westernized countries see the same adverse affects on people from different geographic locations? It is the fluoride causing disease and toxicity, not the country where you live.

The fluoride solution put into water is a waste product from the phosphate fertilizer industry. The phosphate industry is selling people some of their poisonous trash. Our tax money buys it and most people in America drink it. The fluoride solution created when hydrofluorosilicic acid, yes that's spelled correctly, added to tap water may or may not contain additional harmful metals like mercury, aluminum, lead, and other heavy metals, cancer causing agents, and hazardous wastes.

Fluoride is in the water supply. Fluoride is considered a prescription drug by the FDA. Fluoride is not a nutrient. Fluoride is used in people in the United States without permission or informed consent. Fluoridated tap water is used on you and may hurt you without you knowing. Fluoride in water may not help as well as some people think. Do we know for sure? Let us find the answer. Review information from the American Dental Association's position. Then you should decide for yourself and your family.

Does fluoride in water benefit teeth and bones the ways you've been told? Many people don't think so. Many people do. Fluoride can produce toxic effects on teeth, bones, thyroid, and brain functions. Those toxicities are proven.

Fluoride is not conclusively proven to do all the positive things you have been told. Fluoride is proven to do negative things you have heard about but have not been told by the trained appropriate healthcare professional. Why is fluoride still in the water you drink? Make the answer your active choice. Observe a teenager's smile with mottled and badly discolored teeth before you answer. Fluoride's negative effects are worse in male children of African dissent. Excess fluoride is reported to damage the teeth of most children who consume fluoride in addition to drinking water.

Consuming excess fluoride can cause fluorosis in both teeth and bones. Fluorosis is more than simple visual surface effects causing defects on teeth and bones. Fluorosis reduces the strength of the damaged teeth and bones. Make both the affected teeth and bones better. Stop using excess fluoride and damaging other teeth. To lessen the possibility of exposing yourself to excess fluoride drink occasional distilled water. How much and how often are decisions up to you. They are also things changing with life around us. Stop making and using products made with fluoridated water to manufacture other drinkable products.

It's not practical, cost effective, or possible to do research on water now. Small studies applying to some tiny multiple of the needed study population, some unknown number X to protect are still done. Hopefully those come without toxic effects. Some drugs of great benefit have bad effects ranging from minor life system alterations to loss of life. It is too late. Water is and must be everywhere, including from the smallest single live cell to the seagoing mammals called blue whales.

Fluoridated tap water is used to make beverages. There are no drugs without side effects. Since the drug fluorine (or ionic fluoride) is put into water, the assumption is one amount or one dose or one "size" fits everyone. Even if fluoride worked, the dose would be different for a child weighing fifty pounds or less compared to an adult weighing three hundred pounds or more, and every person under, over, and in between. Fluoride in drinking water gives the same fluoride dose to everyone, despite the person's age, weight, liver, kidney function, current health, or disease status. That can't be the right thing to do for everyone.

Fluoride is an endocrine disrupter. By competing for iodide thyroid receptors, fluoride alters thyroid hormones, may cause thyroid problems, and possibly is associated with thyroid cancer. About thirty countries in the world add the fluoride the United States supplies to them for drinking water and use in manufacturing. More people in America drink fluoridated water than the rest of the world. Processed foods and beverages made in countries using fluoridated water contain damaging fluoride. Water fluoridation is a business success that causes human diseases and sometimes failing health problems.

Fluoride is a negatively charged ion, an anion. Two of the common heavy metals that may affect brain function are positively charged ions, or cations. These ions in combination, namely mercury fluoride and aluminum fluoride, may be harmful to the ways your brain is trying to work.

High fluoride consumption through diet and water fluoridation has been shown to reduce intelligence quota (IQ), especially in children. A meta-analysis of fluoride's IQ reduction done by Harvard University researchers reported fluoride's harmful effects on children's IQ. This meta-analysis was flawed. The majority of studies included in the review were conducted in China without some of the control study factors commonly used in the United States. The results could not be extrapolated to the amount of fluoride consumed in the United States.

The study subject's diet was high in fluoride. Drinking water with higher than United States allowable amounts of fluoride. In my opinion, children with a reduced IQ can become adults with a reduced IQ. Fluoride is contained in many products used for babies, children, adults, and physiologically aged adults. Product labels may or may not indicate fluoride contents.

Halogens include chlorine, bromine, iodine, fluorine (fluoride) and more. Chlorine is in the water you drink. The United States puts chlorine in the water to reduce infections. It has done so for more than one hundred years. Chlorination has effectively reduced infections in people. Does chlorinated water still kill bacteria? Some bacteria in water can't be killed by anything. Those bacteria can kill susceptible people, including people with weakened immune systems.

Potassium bromate, a form of the halogen bromine, is used to bleach flour, sugar, and for other uses. Bromine contaminates drinking water. Bromate use is banned in Canada, China, and the European Union. Bromine and bromate are still commonly used in the United States.

As the concentration of fluorine, chlorine, bromine, iodine, organic toxins, metals, drugs, including occasional estrogen-like, androgen or male hormone-like, or feminizing hormone like effects when using hormone pretenders, including herbicides, pesticides, fertilizers, perchlorate, petroleum metabolites, cancer causing agents, discarded substances, toxins, unknowns, and by-products of inevitable human successes and errors rise and fall in drinking water, more people's tastes barely notice. Fewer folks complain. Help make water cleaner.

Tap water regulation authorities often report water as good and pure. Some water systems allow officials to add something to current water supplies if they think they should. Have those additives been proven safe for all people, even sick ones, or kids? Do they always tell you when they add something? Should you trust them? Has anyone anywhere ever let your trust down with water?

What precautions should people have taken before and after water pollution from recent United States hurricanes like Katrina, Irene, Harvey, or Irma? Or from extreme weather occurrences like the typhoon in the Philippines, four feet of snow in a storm in middle Arizona, and other severe weather events? Does everyone get those cautions soon enough? Life will tip over if water keeps tipping the wrong way. In my opinion, it's tipping toward more contamination, new and excess toxicities, adverse human biological effects, sometimes more serious negative effects, and possible disease associated harmful outcomes. This tip will happen sooner and with increasing mortality numbers. How many more people will get sick and die, with or without a vaccine in their arms, or family by their side?

Elements like aluminum, cadmium, lead, mercury, arsenic, and others have "serum levels" or amounts in body water or blood ranging from normal, low, medium, high, or sometimes toxic. My thoughts are these labels are wrong. Those amounts are population-derived abnormal "averages" not established "healthy" normal numbers. There are no normal, or "safe" quantities of these elements in people.

The healthy or normal "blood levels" of those substances is zero. In my experience, none of these elements do anything positive in people. All have damaging effects. Every amount in the bloodstream is higher than it should be. There should be none there. These substances are not the "minerals" you think you need more of but don't get enough of.

Product advertisements are "placed" everywhere so consumers can find needed merchandise, and manufacturers can make sales and profits. Businesses pay to have products appear in movies, radio, television, commercials, on billboards, and other advertising spaces. The unwanted or toxic effects of those products are either not mentioned or are sometimes mentioned faster than you can listen or comprehend. Things thrown away by people or discarded by industries sooner or later temporarily or permanently travel through, arrive in, poison, and pollute water.

Don't believe some forms of advertising. Hype doesn't have to tell the truth or make sense. Seeing and hearing about things makes us want products we didn't want before. Repetition and familiarity creates desire. Seeing and hearing about issues multiple times reprograms our brains.

Water is advertised. Water isn't treated as an essential life component. Water's preparation for sale can increase its contamination, toxicity, cost, and carbon footprint.

Toxic water hurts and kills people. Good water means better life. What once was a whisper is now a scream. Death and sickness are screaming at people from toxic water everywhere. The danger from water must be heard and seen before more bad things happen. Somctimes water kills you slowly. Sometimes water kills you quickly. Either way you're dead.

You should say, "I am the hero, or heroine in this book." These words speak to you and want your actions to speak back to them, so water and people everywhere will benefit.

This book will never be a movie where you see a tsunami of toxic water waste washing over people you know.

There is already a wave of water toxins flowing through and poisoning everyone from the inside. Toxic water may cause children to die faster and live fewer years than their parents. It's already happening in many parts of the world. Small population reductions have happened three years in a row in America. Bad water is pushing adults and their children farther from individual destiny and closer to societal fate.

Drinking any water not distilled exposes you to toxins. Drinking anything made with water not distilled exposes you to more toxins. The better you maintain good hydration, the worse you're doing by drinking bad water, accumulating toxins, and possibly causing or worsening diseases. Most everybody gets sick. There are more children sick now than in all prior generations. There is more cancer in adults than ever before.

Distilled water is the only clean non-toxic water. If you drink eight eight-ounce glasses of tap water a day to maintain water balance, you may be increasing toxicity while improving hydration. You could be killing yourself with toxins but you're not thirsty, and you might feel just fine.

Maintaining proper hydration is not a choice. The younger you are, the worse bad water is for you. People often use distilled water to mix a newborn baby's formula. Toxins in tap water can cause diseases in kids and adults.

There are no toxins in properly made and managed distilled water.

Alzheimer's Disease may also be associated with aluminum toxicity.[210] Associated with doesn't mean it's proven to be caused by aluminum. To be safe, reduce or eliminate your body sources of excess aluminum. Some water you drink contains aluminum.

Aluminum isn't a mineral people use or need. Aluminum serves no function benefiting human physiology, or normal bodily functions. Alone, aluminum may be more toxic to people than mercury by itself. Some investigators report additive or synergistic toxic effects from the combination of aluminum and mercury, which is greater than either metal used alone. The real answer is not yet known. Use your judgment. Your judgment is as good as anyone else's on this subject. There is just no good human observational data available yet.

Aluminum may be toxic to the brain, central nervous system, and bones. It may be as toxic, or more toxic, than some other heavy metals. The problem is we really don't know. Effective comparison between heavy metal toxicity potential has not been evaluated long term in enough people.

Find the causes for current problems before adding more unknowns to the water we drink. Sickness and death are pushed onto us by bad water choices. How many days do you have with toxins collecting in your body from drinking water? How many without? Do better despite your answer.[211]

If you follow recommendations and drink more water, you get more of the natural toxic element arsenic absorbed into your body. If you don't drink more water or other liquids, you may get more dehydrated. If you drink the right amount of water as distilled water, you get neither toxic from arsenic or other chemicals, nor dehydrated. Arsenic contaminates water from an organo-arsenic compound used as an animal feed additive and used to retard certain fungal growths. Arsenic is found in chicken feces. Chicken poop is "harvested" and sold as fertilizer, perpetuating arsenic contamination.

An organo-arsenic compound was added to chicken and pig feeds to prevent diseases such as intestinal parasite infection, increase feed efficiency and promote, animal growth. Arsenic was banned from use for animal feeds by the FDA in 2011. Arsenic is still in some fertilizers, and conclusively in the water supply. Many toxic pesticides contain arsenic and other poisons. Arsenic kills bugs. Arsenic kills people at higher doses. Arsenic exposures, or doses increase as arsenic containing pesticides accumulate in the environment.

During September 2012, it was widely reported on the national news about the dangers of arsenic in the water supply. The article detailed increased arsenic found in brown rice. Some brown rice is used to make baby formula and cereal. Arsenic and other contaminants like cadmium, lead, and mercury, corrupting food eventually gain access to all water sources.

Prolonged exposure to arsenic from water and food has been linked to certain forms of cancer, skin lesions, increased heart disease risk, neurotoxicity, and diabetes. These other toxins cause severe effects in babies. Don't use tap water to make food for newborns. The soy used in baby formulas is genetically modified. The effects of genetic modification on safety or toxicity from soy foods are not entirely known.

The Federal guidelines or controls about arsenic water contamination are sometimes tolerated or ignored by reviewing agencies. Be shocked when your government suggests drinking more water containing arsenic or other poisons is "acceptable." With the human population increasing more rapidly than the numbers of toxicity or death from arsenic, the percentage of people dying from arsenic can go down even as the number of arsenic deaths goes up. Do the math. Prove it to yourself. Is there so much life around we don't notice the death rate from chemical contaminations increasing? How many people are sick or dying from arsenic induced diseases in the water?

Air pollution circles the Earth. Perchlorate falls from the sky and becomes water and human pollution. Pollution sickens everyone and causes various diseases. Air pollution can cause cancer. Thousands of chemicals contaminate the airspace supposed to protect people, other life forms, water, and the planet. Pollutants are in the air we breathe. Death is the only escape from having to breathe.

Neither the air, nor the water are getting cleaner or better. Now we also make the food more contaminated. We have reached strike three.

Lead

Lead additives to gasoline become chemical poisons in air after burning that fuel. Lead then falls from the sky everywhere gasoline burning vehicles are used. By using lead in gasoline, the United States chose to have faster cars, and slower kids. In 2014, approximately half a million kids in the United States are currently affected by lead poisoning.[212] High amounts of lead contaminate soil, water, air, and various root vegetables. Private aircraft fuel puts lead into the sky. That lead falls into the water and onto the land. Lead contaminates roadways from years of lead used in gasoline and fuels. Lead is washed into streams and rivers as rainstorms and floods wash lead from streets and roads.

All people drink water, eat food, and breathe air. Each of those may be contaminated with lead. The water we drink and the air we breathe did not initially have lead contamination. For decades, people put lead into paint, gasoline, and airline fuel. There are still houses with paint containing lead. When children munch those paint chips, they get lead poisoning.

Airline fuel from private planes releases lead into the air, contaminating people. Please take note, lead has been removed from commercial airline fuel. Lead falls onto people who breathe it when they're under aircraft flyways. Flight paths are everywhere. Private airplanes crisscross the skies. Every state and many countries have private aircraft flying to and from them. Lead still falls from the sky.

When private aircraft dump excess fuel before landing, that fuel and the lead it contains contaminates the water supply and land. Lead toxicity reduces mental abilities and causes various mental challenges in people. Lead toxicity can prove fatal.

If you've never seen a private plane in the sky, maybe you're still safe.

PERCHLORATE[213]

Perchlorate is a component of solid rocket fuels, and bombs. Discussions about how much perchlorate it takes to hurt people will never be sensibly resolved. We know how much is too much, but if you're looking for an amount safe in water or food, you won't find a dependable measure. In my opinion, you can't rely on published perchlorate safety guidelines. Everyone is different. We are all individuals and can manifest the same sickness distinctly from each other. Shared or aggregate risk is different for everyone and is always changing, often worsening.

Perchlorate contaminates most people. Everyone gets perchlorate into their bodies by drinking contaminated water and breathing polluted air. In 2004, perchlorate was discovered in the freshwater drinking systems of thirty-five states.[214] Perchlorate contaminates most freshwater drinking systems in the world.

Any amount of perchlorate can be toxic. Perchlorate is classified as a likely cancer-causing agent.[215] There is never a safe "little bit" of cancer in people. Perchlorate is in fireworks. What kid does not remember fireworks smoke smell? I know I do. Smelling fireworks smoke exposes you to perchlorate. Happy Fourth of July! Stay upwind if you watch fireworks. Is your drinking water source downwind from fireworks displays? Many firework displays are held over water. Find out. Be sure.

Perchlorate is found in flares and sparklers. Perchlorate contaminates human breast milk.[216] Perchlorate is everywhere because it gets released into the sky and it doesn't stay there. Knowing what we know about it, we still put more toxic perchlorate into the sky.

Like other chemical causes of cancer, a single amount or toxic quantity of perchlorate doesn't apply. Some people are more susceptible, and some are less. The chemical perchlorate is toxic, but no one knows how bad it is for each individual person. More perchlorate is put into the sky with every war. This world is always at war.

Missile and nuclear test sites surround parts of the Colorado River. Water from the Colorado River is used as drinking water and irrigation water in Mexico, Southern California, Utah, New Mexico, Nevada, and Arizona. Perchlorate, radiation, and other toxins will contaminate the water supplied by the Colorado River longer than anyone realizes. It is estimated about one-third of people in the United States drink water exceeding limits for radiation as reported by the EWG in 2018.[217] For many years, the diminished Colorado River did not flow into the Gulf of California. It does now, thanks to a treaty with Mexico. It may not in the future. Treaties change. Ask native people where you live.

Perchlorate contaminates

1. Water supplies
2. Milk, milk products, and organic milk
3. Breast milk samples
4. The Colorado, Mississippi, Ohio, Delaware, and all river basins in the United States
5. Foods, including organic foods
6. Various vegetables and fruits
7. Meat and meat products
8. Fish, both freshwater and saltwater.

All the major river systems in the United States were polluted more than fifty years ago. They're worse now.

The philosopher H.L. Mencken said, "It's a shame when God limited man's intelligence, he didn't also limit man's stupidity." Stop building rockets and bombs using perchlorate.

The term vitamin is a contraction of the words "vital mineral." Early "disease cures" were from vitamins. The diseases cured, like scurvy that was cured with vitamin C, were redefined as vitamin deficiencies. The cures weren't called drugs but were recognized as "vital minerals" or substances people get from eating foods. Most drugs treat disease symptoms, not causes.

Human intake of minerals is often misunderstood. Minerals don't come from the common water you drink. Routine minerals come from foods you eat. Trace minerals and elements come from foods and specialty salts. People absorb minerals and trace elements from food, not from water.

People get more minerals from eating one apple than all the water they drink in a week.

Some statements about minerals lead you to believe drinking water is packed with minerals. Water is not loaded with minerals. Drinking water contains few minerals in low amounts. Most people get all the minerals needed through their daily food consumption. Minerals help people perform functions for growth, maintenance of body cells, and health. These actions occur without many of us consciously thinking of minerals or intentionally consuming them.

If your diet is normal or "average," as variable as that may be, you rarely need to consume extra minerals. Eat a little everything, not too much of anything, and drink lots of water. You need to constantly rediscover and consume your normal diet and needed mineral intake from foods as you age. There's no evidence people absorb significant mineral amounts from drinking water.

Dr. Andrew Weil has written "by one manufacturer's estimate, you would have to drink 676 eight-ounce glasses of tap water in Boston to reach the RDA (recommended daily allowance) for calcium." Need Dr. Weil or I say more about minerals in water?

The time for caution concerning toxins contaminating drinking water from ground sources and taps is now. Numerous toxins are here today. Many are dangerous to people.

To get the essential mineral cobalt in their diet, vegetarians and vegans occasionally need supplementation with vitamin B12. One milligram, or 1,000 micrograms, of sublingual vitamin B12, cyano-cobal-amin or methyl-cobal-amin is often as much as they need. Vitamin B12 aids in making red blood cells and repairs nerves. Omnivores are people who eat all foods. Some vitamin B12 is normally furnished to them by eating meat.

Hippocrates said, "Let thy food be thy medicine, and thy medicine be thy food." Hippocrates didn't know people were going to ruin all the water, and most of the food with toxic water and additional impurities from the air we breathe and the soil used to grow foods.

If the mineral issue concerns you, eat a variety of foods. Eating food supplies all the minerals you need. There are few mineral waters on the world's markets. Many mineral waters don't contain minerals in sufficient amounts to make a difference in mineral intake.

In America, if a mineral water contained a therapeutic amount of minerals, the FDA would control it. It would be marketed as a drug with regulations and controls. We would have to drink gallons of water to get any mineral intake. Water, not minerals, moves easily through cell membranes into human cells.

Some articles describe "Filtration methods removing bad minerals and leaving good minerals behind to perform beneficial functions in people." That combination of good and bad simultaneous biological filtration methods simply doesn't exist.

Make distilled water yourself and put it into clean glass containers to be sure of its purity and safety. Reading this manuscript can improve the quality of your life and the lives of people you care about by changing the waters you use.

Take one long slow mouthful of distilled water to taste it. Distilled water has no taste. You'll be shocked by what you don't taste. After drinking distilled water, take a sip of tap water and taste what you no longer want, need, or miss. Spit out the tap water. You already consume too much toxic awful tasting water whether you know it or not.

Getting, bottling, and selling water is one of the largest industries in the world. It is the third largest after the hydrocarbon and electricity businesses. All three industries create pollution, some of which ends up as more toxins corrupting water sources. Plastic is the most common solid pollutant in water and the world. In some water there's more waste than life. Globally more than 80 percent of tap water samples tested contained plastic particles. In the United States, more than ninety percent of tap waters tested contained plastic substances. The majority of the world's water systems are contaminated with microplastic particles. Without knowing, most people drink them. Most filtration systems can't or don't remove plastic micro particles.

Distilled water is the only pure uncontaminated water left to drink. All other water contains toxins, at least tiny amounts. Distilled water is cheap. It costs about eighty cents a gallon to buy. You can pay that now or pay much more later when you're sick from water toxins and trying to get better. The cost to your health from drinking tap water and the diseases it causes is very high.

Eventually, people and companies will tell the truth about the water and drinks they bottle and sell. They'll admit there are toxins in tap water. Many beverages we buy and drink are made from tap water. Some people will tell you they don't often drink the beverages they make. They'll tell us they filter their water and it's aseptic and secure. Some do, but it's not either. They can't prove their water reduces long- or short-term health problems in people. They'll blame something else for the troubles. They'll claim their water produces better health. Those opinions are not scientifically objectively proven or disproven.

Some smart, fancy, or "green" waters cost more than ten dollars a gallon, more than a couple gallons of gasoline. These waters lack reproducible proof about their long-term health benefits for people. Evaluate the facts before you accept their claims as true. Most of those waters are sold in harmful plastic containers.

In my opinion plastic wraps are toxic. They transfer plasticizers, phthalates, BPA, and other harmful chemicals into water, food, and people. Phthalates and plasticizers contaminate water, food products, and children's toys.

Some people say phthalates are toxic. The industry making phthalates says they're safe. Phthalates are associated with, not proven to cause, possible reduction of intelligence, and increased cancers in children. Children drink, eat, breathe, and play with phthalates and plastics.

Distilled water is proven to not contain the same toxins as other types of water. The majority of childhood leukemia is preventable. Some is caused by toxins in water. Distillation removes all toxins.

Cancer currently affects one-third of people in the United States. Cancer is the second most common cause of death. Some people say it will soon become the first, but not everyone believes that. Cancer mortality has tripled in the last one hundred years. The World Health Organization estimates the cancer rate will more than double in the next twenty years.[218]

Buy, store, and drink distilled water in and from unpainted glass containers. If you buy distilled water in plastic gallon jugs, pour the water into a clean glass gallon bottle as soon as possible. Just like wine, uncap or uncork the bottle, cover it, and let it "breathe," or eliminate some toxic gases for a while before it is consumed. Letting water breathe allows some but not all volatile organic compounds, toxins, and plasticizers to escape into the air. Heating and/or cooling water temperature may change the solubility and amount of toxins staying in the water, aiding in efficacy, or conversely promoting toxin elimination. Stay focused.

The effects of poison water created from combinations of toxins aren't well known or studied. Often these water contaminant combinations only have chemical names and little negative toxic or long-term positive safety and effectiveness information clearly identified and confirmed. Much information exists describing positive information about water's benefits, and negative risks from contaminants in water, their immediate problems rarely, and short, or long-term risks of toxicity.

These chemicals may be either good and helpful, bad and toxic, or both. I think the combination of toxins is likely more toxic than either alone. Neither their safety nor toxicity are clearly identified. Could human water balance changes, either positive or negative, be ahead as these chemicals continue to be used. Water is full of perilous mixtures. Neither harm nor safety from plasticizers will ever be clearly proven to my satisfaction.

Public tap water can be unhealthy. Consider Flint, Michigan in 2015-2016. The problems weren't just caused by contaminated water and old lead pipes. Some troubles were caused by people we should be able to trust. They made bad decisions. Some were just "following orders." Some were arrested for the problems they created, and others were charged with involuntary manslaughter. Legal evaluation is continuing as of 2022.

Flint's children made sick by the water will be sick longer than it takes to fix the problem. Who's fixing those children? The government didn't make the good water bad. They made bad water worse. Make Flint's water better than anywhere else. Use non-toxic healthy distilled water to fix the water problems.

Flint's problems are happening elsewhere in America and all over the world. There are thousands more communities with water toxicity as bad as Flint's. East Chicago, Indiana was in the news for water contamination. Their problems are as bad or worse than those in Flint, Michigan. Keep listening. You will hear about many others.

There have been reports listing lead concentrations in drinking water hundreds of times higher than the EPA's recommended 15 parts per billion limit.

People in cities are sometimes advised to boil water to kill biological contaminants causing infection. In June of 2012, news agencies in Camden, New Jersey advised residents to boil water to reduce infection risk. How many people didn't see or hear that warning? What happened to them?

If tap water smells bad, don't drink it. If tap water smells good, don't drink it. If tap water smells at all, don't drink it. Clean non-toxic tap water doesn't have odor.

People in Montgomery County, Pennsylvania, were advised to boil water before use. This was also reported on their local news. The contaminant was reported to be a mineral. Boiling water does nothing to reduce mineral concentrations. Should we trust the instructions we're given if they aren't understood by those giving them to us or if they don't make sense?

Montgomery and Bucks counties in Pennsylvania were warned again about water contamination with cancer causing chemicals in May of 2016. The firefighting foaming chemical perfluorooctanoic acid (PFOA) and others were reported as contaminants in various water sources. If you live in either county, don't drink local water or beverages made in those locations. You risk damaging your good health if you do drink them. Do your best.

Just after my most recent years living in Pennsylvania, I dropped out of cultural life. I ate food I raised and drank water from my well. My cabin had a fridge and a stove, but no electricity to plug into. Alone on a mountaintop with no other people around and no connecting road, I saw no one for many years. Happy had a new definition. I saw nothing but trees and wild animals for miles and miles. One stray cow I nicknamed Bessie visited about once a year before just moving on. Not sure where she escaped from, or went back or on to. Still a mystery. There was no one to ask and she wasn't telling.

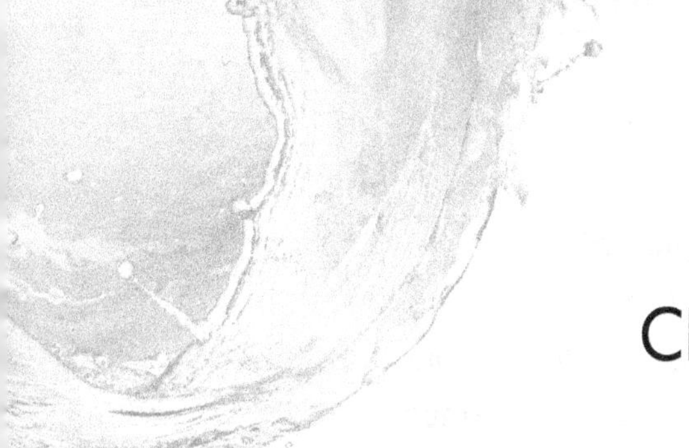

CHAPTER 9

Niagara Falls has frozen many times. Winter 2014 was one of those icy occasions. A huge section of the river and falls was chilled into silence. Most of the water previously plunging into thunder was quiet and unmoving. The silence seemed everlasting. The roar of Niagara Falls normally drowns out all close-by sounds. The sound from Niagara Falls may normally silence the truth about danger from chemical contamination, and human harm from possible toxins. Niagara Falls is angry now because of water toxins and contaminants from Love Canal.

Hooker Chemical Company buried twenty thousand tons of toxic waste in the Love Canal area. One study showed a birth defect rate of 56 percent in children born there. It was not 5.6 nor 0.56 percent. Both of those low numbers would have been too high. Fifty six percent affects fifty-six of every one hundred children born there after the improper waste disposal.

Fifty six percent means more than one of every two, or slightly over half the babies born there during that time had a birth defect displayed in many different systems.

How can people poison one of the world's natural water wonders and the people who live in the neighborhoods surrounding the falls? People and companies who poisoned the environment made money from this situation. Why has Love Canal not been totally cleaned? We'll need that water and land again someday to live. Love Canal is a superfund site.

Children have been hurt and killed. Kids and adults died from the water. Money was spent to identify and study the toxins in Love Canal, the Niagara River, and at Niagara Falls. Enough money was not used to completely clean the water or contaminated land.

Funding for the national superfund cleanup expired in 1995. Some superfund sites did not get completely cleaned or go away. Superfund funding should be a constant budget item. Superfund site numbers are increasing.

If the Love Canal poisons found in humans were in fish, people wouldn't be allowed to catch and eat those fish. Those fish got toxins from living in contaminated water and eating other fish and contaminated foods.

I think some people in the least developed places may have some of the world's cleanest water.

"Millions of Americans drink unsafe water because of chemical or bacterial contamination."[219]

One-third of American waterways including lakes, rivers, and bays, are unsuitable for swimming, or for catching and eating fish. "Drinking water in the United States contains over twenty-one hundred substances, some causing cancer, problems with human cells, and nerve disorders."

More than sixty million Americans drink water polluted with feces, urine, radiation, and other contaminants. This causes infections killing nearly three thousand people each year. "There are almost four hundred thousand cases of water borne illness in the United States each year. Nearly all of these infections are minor and most infected people recover with no permanent effects or problems."

Twenty years ago, the EPA estimated that fifty-three million people were exposed to toxins or infectious agents by the water they drink.

The Environmental Working Group (EWG) found herbicides in every Midwestern city water supply they tested.[220] The EWG estimates that twenty-five million Americans are drinking herbicide-laden water. One city tested had thirty-four times the amount recommended by federal governmental standards of the herbicide, cyanazine. Cancer and birth defects happen to animals fed these herbicides.

Water treatment facilities were not built to remove toxic substances from drinking water.

People don't think they live near the water they pollute while putting themselves and their kids at risk. We are all in danger. Water is supplied to communities through new and old pipes made of metals like lead and copper, or synthetic plastics like PVC. In my opinion, each kind of pipe adds contaminants to water. Pollution problems from piping are widespread. Benton Harbor Michigan was in the

TV news this morning reporting lead toxicity in drinking water coming from lead pipes. Those problems add to the toxic effects of the water we drink.

If you have a system removing toxins from water, but the new cleaned water produced is delivered through old toxic pipes, that doesn't solve the water contamination problems.

When people test water, and report it's completely uncontaminated, that may mean they haven't tested it for the correct substances. Extensive and proper testing always shows some contamination of any water that hasn't been distilled. Steam distilled water put into clean non-toxic glass bottles is the cleanest water possible. Distilled water is the only contaminant free water.

Clean water regulations may be relaxed to accommodate increased toxin issues. Water system changes for political and other reasons may increase the water supply but may not be providing uncontaminated water. Sanitary water needs increase while clean water supplies decrease. People get sicker and more toxic as water gets "dirtier." Allowable and approved doesn't always mean safe or non-toxic.

People began poisoning water as soon as they began using land.

The enhanced photo-shopped water seen in movies and on television can be quite beautiful and attractive. If only our water were that clear and clean anywhere it occurred naturally. Unless distilled, water looking that good on Earth cannot be. Don't believe the hype, most advertising, or propaganda about water. All life gets contaminated from toxic water. Cleaner water may be less toxic but it's still somewhat toxic. Many toxin removal processes are not even 50 percent efficient. Except for steam distillation, toxin elimination processes always leave some poisons in the water. The effects from most toxins, especially toxin combinations, are not completely discovered or totally understood.

Some water cleaning processes add different compounds to the water. Are those substances harmful? How much is safe to drink? How much is toxic to children or sick people? The substances added to water haven't been adequately studied for the toxicities they might produce.[221] Some water that looks "cleaner" may be "dirtier" and more toxic.

Cooking your food reduces bacterial contamination. Often when eating or drinking, our only choice appears to be no choice. Temporary "fasting" is a better choice than drinking toxins or eating bad food.[222] We can't escape from pollution. We pay with our lives for dirtying, then drinking unclean water.

The business of America is doing business. In the pursuit of power, what people make best are profits and pollution. Business leaves toxic environmental footprints.[223] More than sixteen billion feet on the planet make footprints. Human "footprints" pollute the water we use and drink. Business is messing with the good, positive, natural outcomes routinely produced by most of the world's water.

Mess with the water, and chaotic problems soon follow. Most is less than half, but probably more than just some, according to the eye. More than one out of every ten people you know experience disappointment or "Hot Water"-pun intended-dealing with life. Pollution of water and food could be the ultimate cause of death for life forms. If not a cause, it could for sure be a catalyst.

I believe some pollutant combinations create new toxicities in people. The new pollutant may have improved solubility, penetrate more tissues, and cause more severe adverse effects. When you add one pollutant to another, you never get a combination safer than the individual parts and you probably get a more toxic mix. We don't know about the toxic effects from pollutant combinations. A single found pollutant in people is rarely alone.

CHAPTER 10

Article Titles, Information, and Misinformation from the Internet. Some people can't tell which is which sometimes. Whether you, or me, or anyone else, everyone can, and everybody should decide first what's real. I follow-up most publishers, but I can't assure the accuracy of one hundred percent of each sample's content and printed output. I have reason to trust every source herein or I would not have included them in this document. Think first, then decide for yourself (references and titles between quotes can be found on the internet). If you don't have the ability to get on the internet from where you live, please get to a town with a public library. Libraries often have internet connections, devices to connect you, and people to help you do that. Some friends will even print your results.

The statements in quotation marks "..." in this chapter can be found on the Internet when you search for information on water and on some of the agencies regulating water systems in the United States and around the world.

Words not between quotation marks are from me for emphasis and to provide context as you read and research this topic. These statements exist now on the Internet. More information will be available by the time it takes you to read this book, however you read this book, and how often you choose or need to reread.

1. "Are there drugs in our water supply?"

Tiny amounts of some drugs enter your body with the water you drink. Water in bottles begins with tap water that's polluted with medicines. Read chapter 8 again for more detailed information on this subject.

Drugs are not present in sufficient quantity to exert primary effects. No worries. However, in some very few people, medicines may be present in high enough quantity to occasionally cause adverse or allergic reactions. This may or may not ever happen.

2. "The United States bolsters chemical restrictions for water."

"There are a range of chemicals that have become prevalent in products, our water, and our bodies in the last 50 years," the former EPA administrator, Lisa P. Jackson said in a speech.

"The Clean Water Act has been violated more than a half million times we know about. There is a history of this agency making big announcements, and then changing very little," one EPA regulator said.

In addition to these statements by members of the EPA, please see chapter 4 for more information on the CDC's view on water and its contamination.

3. "Rulings restrict Clean Water Act foiling EPA."

"Thousands of the nation's largest water polluters are outside the Clean Water Act's reach, because the Supreme Court has left uncertain which waterways are protected by the law, according to interviews with regulators."

That surprised me when I read it. Some "companies having spilled oil, carcinogens, and dangerous bacteria into lakes, rivers, and other waters aren't being prosecuted."

"Environmental Protection Agency regulators working on those cases estimate more than 1,500 major pollution investigations have been discontinued or shelved in the last four years. About 117 million Americans get their drinking water from sources fed by waters vulnerable to exclusion from the Clean Water Act, according to EPA reports."

"But midlevel EPA officials said that internal studies indicated that as many as 45 percent of major polluters might be either outside regulatory reach or in areas where proving jurisdiction is overwhelmingly difficult."

Additionally, political powers are afoot with the changing of each administration. Strengthening or weakening the EPA becomes a game of political tug-of-war. In a model situation, we should be able to trust leaders to be keep this country's water and citizens safe.

4. "Clean water laws are neglected, at a cost in suffering."

"J. H-M knows not to drink the tap water in her home near Charleston, West Virginia."

"Her entire family tries to avoid any contact with the water. Her youngest son has scabs on his arms, legs, and chest where the bathwater - polluted with lead, nickel and other metals - caused painful rashes."

"Tests show that their tap water contains arsenic, barium, lead, manganese and other chemicals at concentrations federal regulators say could contribute to cancer and damage the kidneys and nervous system."

"How can we get digital cable and Internet in our homes, but not clean water?"

"...research shows that an estimated one in ten Americans have been exposed to drinking water that contains dangerous chemicals or fails to meet a federal health benchmark in other ways."

In my opinion, I do not believe the investigation was thorough enough to conclusively determine contaminated water was the cause of her son's maladies. If this can be definitely proved, then my hope is the situation is resolved, her son experiences a full recovery, and the water can become clean. Until then, I remain skeptical, not agnostic, about this report. I need more information to be convinced with any conclusion.

5. "Saving United States water and sewer systems would be costly."

"In Washington (District of Columbia) alone, there is a pipe break every day, on average ... causing untreated sewage to flow into the Potomac and Anacostia Rivers."

" ... and so each year, thousands of ruptures damage streets and homes and cause dangerous pollutants to seep into drinking water supplies."

The crumbling water infrastructure in the United States has become a political battle cry for the current administration. Whether or not our leaders will act to improve the foundation and protect citizens further remains to be seen. In the meantime, the decay of the sewer systems of this country continues to endanger our already contaminated water, water sources, and personal health.

6. "Toxins in our drinking water."

"Water is the miracle elixir that can sustain us if we never drink anything else." That statement was true before people ruined the water. It can be true again if the world's water quality and safety can be restored.

"Some people actually drink enough water. However, they may not be drinking the right water." There are more than eighty regulated contaminants and more unregulated toxins, like the rocket fuel component perchlorate present in most tap water.

Those chemicals cause danger and overburden our immune system parts, kidneys, liver, skin, colon, and other systems.

7. "Documentary examines how toxic water at the nation's largest marine base damaged lives."

"MP did not believe the rumors about a place called Baby Heaven until he visited a Jacksonville, North Carolina graveyard and wandered into a section where newborns were laid to rest. Surrounded by hundreds of tiny marble headstones, he started to cry."

"The Marine Corps at Camp Lejeune routinely dumped fluids containing dangerous chemicals, which leached into groundwater, and eventually contaminated drinking water wells."

"For decades, buried tanks leaked fuel allowing known carcinogens like benzene into the ground nearby."

"But Camp Lejeune failed to study the health risks of its water after toxic compounds were discovered in the early 1980's and did not notify Marines and their families. Up to a million people ... relied on the water to drink and bathe."

"The Marine Corps has said it was not aware of the contaminants until the mid-1980's and contacting the seven hundred fifty thousand to one million military personnel who lived at Camp Lejeune during those decades is too large an undertaking."

It is well documented the United States has one of the highest infant mortality rates of the developed countries. Babies die. Did all of these babies die from contaminants in the water they consumed either through mother's milk, infant formulas, and/or bathing? No one may ever know the answer and neither do I.

8. "That tap water may be legal but may be unhealthy."

"The 35-year-old federal law regulating tap water is so out of date that the water Americans drink can pose serious health risks - and still be legal."

" ... Not one chemical has been added to the list of those regulated by the safe drinking water act since 2000."

"Other recent studies have found even some chemicals regulated by that law pose risks at much smaller concentrations than previously known."

"However, many of the act's standards for those chemicals have not been updated since the 1980's, and some remain essentially unchanged since the law was passed in 1974."

"From drinking contaminated water, ... millions of Americans become sick each year with maladies from upset stomachs to cancer and birth defects."

9. "Millions in the United States drink dirty water, records show."

"The Safe Drinking Water Act requires communities to deliver safe tap water to local residents. But since 2004, the water provided to more than 49 million people has contained illegal concentrations of chemicals like arsenic or radioactive substances like uranium, as well as dangerous bacteria often found in sewage."

"Studies indicate that drinking water contaminants are linked to millions of instances of illness within the United States each year."

"It is unclear precisely how many American illnesses are linked to contaminated drinking water. Many of the most dangerous contaminants regulated by the Safe Drinking Water Act have been tied to diseases like cancer, taking years to develop."

10. "Congress releases report on toxic chemicals used in fracking."

Fracking extracts natural gas and oil from shale beneath the Earth's surface. Communities are concerned about the leaking of chemicals into aquifers, rivers, oceans, streams, and the atmosphere. Benzene, toluene, ethylbenzene, and xylene are common chemicals fracking uses. Many more chemicals are used.

Many are Volatile Organic Chemicals (VOC's) that may or may not have to be disclosed to the public. Fracking may use a mixture of more than seven hundred fifty chemicals and millions of gallons of water per frack. The solution is recovered, the gas or oil separated, sold or used, and the remaining fluid then disposed of with or without treatment. Disposal is often in wells producing drinking water. People drink that water without knowing what chemicals contaminate it.

11. "Fracking update: Methane in drinking water near drilling sites."

A study done in Pennsylvania and New York, within a half mile of a gas-drilling site, found wells contaminated with methane. Half of the sixty-eight study sites were contaminated.

Recent fracking probes have uncovered dozens of chemical toxins, and carcinogens in drinking water. Your water may contain enough methane to be a fire hazard. What? Water can catch fire. Some gas company officials say methane is safe to drink.

12. Report: "Carcinogens injected into hydrofracked wells."

"Chemicals were injected into wells."

Hundreds of millions of gallons of hazardous and cancer-causing chemicals have been used in the fracking process. Twenty-nine chemicals are known or suspected causes of human cancers. These chemicals are "regulated" by the Safe Drinking Water Act or listed as hazardous air pollutants under the Clean Air Act.

13. "Chesapeake Bay"

The drainage basin for the Chesapeake Bay covers over sixty-four thousand square miles in six states. Dead zones with depleted oxygen concentrations kill about seventy-five thousand tons of seafood or food for seafood, every year.

Blue crab and oyster populations have dramatically diminished. Large fish kills have happened. Numerous causes for reduced oxygen and diminished sea life include algae blooms, fertilizer and pesticide runoff from farm fields, sewage, and other toxic issues. The Chesapeake Bay, the largest estuary in the United States, is only a generation away from being declared dead.

14. "Do we receive our minerals from drinking water?"

" ... A big misconception is that we obtain enough minerals from our drinking water."

"In order to receive enough minerals for our bodies, we would need to drink a full bathtub amount of water every day!"

We get minerals from the variety of foods we eat. There are small quantities, but no significant amounts of minerals in water. We don't efficiently absorb minerals from anything but food.

15. <u>"Methyl tertiary butyl ether in drinking water"</u>

MTBE or methyl tertiary butyl ether is a suspected carcinogen. The Environmental Working Group found that toxin in more than fifteen hundred water supplies from twenty-eight states. An estimated fifteen to twenty million people in the United States are exposed to tap water containing MTBE.

16. <u>"Study finds 232 toxic chemicals in new born babies, including BPA."</u>

This study was done by the Environmental Working Group. They analyzed umbilical cord blood from babies and pregnant women. Many chemicals were found in the blood. Whether toxic or not, don't expose newborns or others to a chemical not proven safe.

The study found over two hundred toxins. Many more could have been found if others had been named, identified, and looked for. Newborn babies start life exposed to numerous foreign substances. When infants go out into the world, they are bombarded with thousands more chemicals. It's not easy to live in America now and not acquire additional chemical substances.

The best way to start detoxifying, or detoxing a baby is to stop them from collecting more toxins from foods or drinks, including water. Stop kids by stopping yourself. Kids do what they see. Children may download behaviors observed from the adults they admire.

17. "Warning: Nutrasweet® is a neurotoxin."

NutraSweet® is a brand name for the chemical aspartame. Aspartame is also available as Equal®, Spoonful®, and other brand names. Aspartame has thousands of reported adverse effects in ninety-two "documented" categories. Nutrasweet® is associated with a dozen diseases thought to be triggered by it.

Nutrasweet® is the substance most often commented on to the FDA. Those comments are mostly about problems, not thanks or praise. Controversy surrounds the use of NutraSweet®. Various liquids, including some drinking waters, contain NutraSweet®.

> Like, not identical to, but also seriously unlike NutraSweet®, so also maybe as much unlike Nutrasweet®, Splenda® or sucralose, is an artificial sweetener. After human consumption Splenda® is eliminated unchanged in the urine. Splenda® contaminates the water table.
>
> The Los Angeles Times reported that the Grand River in Ontario, Canada has been found contaminated with saccharin, Splenda®, and other artificial sweeteners. That river feeds "fresh" drinking water systems in several states in the United States. Has Splenda's safety been objectively proven without question or doubt? Splenda® is reported to be associated with many toxic effects including possible cancer in animals.

Through complex chemical and pharmacological mechanisms, this non-caloric sweetener may cause weight gain, despite containing no significant calories. Find out whether any of these reported effects are real, their incidence, and any treatment before the artificial sweetener is used more. Nutrasweet® is contained in thousands of products. Many "diet" beverages contain toxins from tap water, and Nutrasweet®. Many "diet" foods contain Nutrasweet® and other artificial sweeteners.

18. "Polybrominated diphenyl ethers are PBDEs"

The PBDE flame retardant chemicals are everywhere. They are associated with many problems and some cancers. They were used on numerous commercial products. Chemicals that can prevent fire from happening should be thoroughly studied and objectively evaluated before use in people.

19. "Where are America's ten most polluted rivers?"

The short answer is "everywhere." Rivers and other waters in the United States are polluted. We're all downstream from one of them. Over fifteen years, the CDC documented more than two hundred fifty disease outbreaks and nearly a half million cases of illness from drinking water in the United States.

20. "Are you eating, drinking, and breathing Monsanto's new agent orange?"

Numerous articles report on finding RoundUp®, AKA glyphosate in the environment. RoundUp® may be in the water you drink and on the foods you eat. The frequency of detection ranged from sixty to one hundred percent in both air and rain samples. RoundUp® herbicide is commonly found in rain and streams in the Mississippi River and other major United States river basins. There are many more articles about RoundUp® in America and around the world.

21. "Water, the essential nutrient" was written by Andrew Weil, MD.

"Water is a basic necessity, needed to maintain a healthy body, a clear mind, and a good balance within your tissues. About 60 percent of your body is water, and you must constantly replenish the supply, as it's used continuously in the processes of life."

22. "Water watchers" was written by Darcy Kiefel

In the Philippines, Serafin Billones ... "recognized that his survival depended on the ebb and flow of water ... the rushing waters of the river could be heard for miles. Today the silent water barely covers the sand on the shore."

23. "Coal chemical spills in West Virginia, leaving 300,000 without drinking water." was written by John Upton and published in Grist Magazine on January 10, 2014.

Between two and five thousand gallons of a toxic Methanol compound leaked from a holding tank and contaminated a drinking water source.

The contamination caused declarations of State and Federal emergencies in Charleston, West Virginia. Government officials asked people to not use the water for anything except flushing the toilet and firefighting. Restaurants, hospitals, schools, and other institutions were affected. Water used for firefighting and the toxins it contains enters and, consequently, contaminates the water table.

The methanol compound leak was confirmed by state officials after a report of odor in the water by citizens. Contaminated water was used before discovery. This contamination occurred in addition to the metal toxicity already there.

24. "Troubled waters, rising demand for freshwater confronts limited supply." was written by Sandra Postel.

Ms. Postel writes "Less than 3 percent of Earth's water is fresh, and two thirds of that's locked up in glaciers and ice caps." Some freshwater ice caps and glaciers have melted since then. The ice melted. Fresh water mixed with saltwater. There is less freshwater now. The freshwater crisis is worse.

25. "You absorb water when you shower."

Some people say you absorb one gallon of water during a ten-minute shower. Prove or disprove that yourself. One gallon of water weighs eight pounds. A pint weighs roughly a pound the whole world around. Weigh yourself before and after showering. If you don't gain eight pounds or more from a shower you haven't absorbed one gallon of water. If you had absorbed one gallon of water, you would be peeing that out over the next couple of hours.

Statements like these are easy to find on the Internet. I hope they compel you to read more, believe what you can prove, and act in the best interest protecting your health. Need I write more? The water system in our country is in shambles. Improvements have been made but much contamination chaos continues.

The world's dwindling safe water supplies must motivate every level to change now. Drinking distilled water and beverages made with distilled water will help stop a major source of toxin body-load absorbed from the tap water you drink. Everything made with distilled water tastes better. You don't taste the muted and diluted metallic flavors of mercury, aluminum, lead, other metals, contaminants, toxins, microorganisms, and bacteria in tap water. Distill your own water, and when possible use it to make everything you intend to drink. Eighty percent of the Earth's population drinks coffee. Using distilled water to make coffee is an easy recommendation to make. Let billions of individual tastes lead the way to your frequently varying happy choices.

Those poisons surround you. They are in the air you breathe, water you drink, and food you eat. Some toxins can be reduced or eliminated by drinking more clean water. Others, especially heavy metals and poisons, are not easy to remove.

The United States Clean Water Act was passed by congress in 1972. Published information from 2016 indicates the clean water act is still not fully implemented. As water pollution's legal limits are reached and exceeded, laws may be adjusted, relaxed, or forgiven to benefit the polluter. Water pollution gets worse for many reasons, including air pollution. Pollutants from the air constantly fall into the water. The Clean Air Act from the 1970's is often ignored or not enforced. As of this writing, the Clean Water Act is possibly due to be repealed or changed again. Changes in regulation are well intentioned for the benefits of the population. I want to believe regulators begin by wanting to improve water quality and delivery for the world's population parts they serve. Societies around the globe are trying to do the right things for water. How often do they achieve or fail to achieve that goal?

Only rarely are companies penalized for violations of the Clean Air or Clean Water acts. Penalties may only cost money. Those fines rarely result in changes to polluting business practices. Fines may fuel the ineffective bureaucratic governmental anti-pollution machine without making real change. The ecosystem contains chemicals which cause cancer. Every bit of smoke ever produced went into the atmosphere, the only sky we will ever have.

Bad food quality worsens water value. Organic and non-organic USDA additives move from food into the water. Organic farmers aren't told not to spray, they are told what and when they can spray. The toxins from water get into people. As bad or good as food products are now, including organic foods, make sure they don't get worse in future years. The quality of organic foods has been reduced with changing regulations.

Polluted water is used to grow fruits, vegetables, animals, and fish. Water is made worse by adding the chemicals used to grow food to the water table. The life of fish is confined to the water essence surrounding them. The United States Department of Agriculture (USDA) approves twenty additives used for organic produce that aren't organic. That list includes two meat products. There are no organic animals. The product›s organic label doesn›t tell you that. Those additions and other herbicides, pesticides, fertilizers, and toxic chemicals spoil the water we drink and the food we eat.

The organic label is applied to foods 90 percent organic. Apply whatever trusted pretend pseudo-positive adjective you like to food where questions persist unanswered. Ten percent toxic makes water more toxic. Ten percent trash or poison is still trash or poison. Don't settle for problems that should be resolved. Perfect clean water began life. Dirty water is ending life. Will more water toxicities end more lives too soon?

Eating organic fruits and vegetables reduces dietary toxin exposure from foods. Drinking and eating right makes you feel better psychologically and physically soon after starting. Drinking juice you make from organic fruits or vegetables and distilled water, supplements minerals from fruits and vegetables while you minimize toxin intake. There are more minerals in those fruits and vegetables than in any water you drink. Drinking from glass improves the water's taste and helps the detoxification process by reducing toxin accumulation. Success helps build greater success. Good feelings make you want more good feelings.

Pure water, clean air, and uncontaminated foods are important for good health and long life. We could live longer if we drank non-toxic water and ate cleaner food. Don't drink or eat anything with an ingredient label. The longer the label the greater the chance for toxic accumulations in some cases. Some labels list more than one hundred artificial ingredients contained in the product. That product often is a food.

One hundred years ago, we didn't live as long but spent only a shorter time dying. Now we live longer but spend more time dying.[224] Our lifespan is longer, but our health span is shorter. If you spend your life dying, are you living or just alive? The prevalence of cancer has always been increasing. Although much progress has been made, the severity and overall cancer mortality rates continue to rise.

Some pollution situations are improving but other pollution problems continue to increase. Diseases caused by the toxic environment worsen while waiting for legal action resolution. Forty years is enough time to settle issues and begin to fully implement laws already passed.

The title "Clean Water Act" makes people think and believe there's governmental oversight reducing water pollution and building cleaner water. Most water problems aren't addressed by the Clean Water Act, and water pollution many places worsens.

If mortality from water diseases decreased under the Clean Water Act, you could be less concerned about the purity of water. You would know the rumors, both clean and positive or dirty and dangerous, about the water you drink. That isn't what's happening.

In my experience, minimally one in ten children is sick now. In twenty years, disease prevalence may be one in two children. Half a generation of children may be getting sick. Enforce the climate codes. Fix the pollution problems. Clean the water. Older wisdom used to say water isn't clean until it runs over many rocks countless times. That means many complementary methods must be used to help clean the water. Save the future.

The sick care system doesn't prevent diseases and "acute or now" water pollution problems, crises, or chronic ongoing problems with water delivery, use, and subsequent safety. People are sometimes told these problems should be accepted as part of modern life. The roles of various types of pollution in these diseases continues to be debated with little wisdom realized and few issues settled.

We walk, run, swim, or wear ribbons for cures as the incidence of some diseases worsens. Most disease causes are not identified or prevented. Most electricity in many world locations is produced by old junky coal-burning systems.[225] Despite what you hear, many of those systems haven't been fixed. Mercury from coal burning worsens some diseases and may be implicated in causing others.

To reduce air pollution from these plants, some air "scrubbers" are used to clean the process of electricity generation by reducing other air pollutants. This doesn't reduce overall pollution. The multiple pollutants recovered from the air are captured in water or mixed with soil– best cases; and spread on land. That water is returned to the water table or added to land that is somehow pre-water table, perhaps layered with various water-resistant products. To breathe fewer toxins, we drink more poisons. Dilution is still not a solution.

If water and other pollution stopped immediately, it would take several average human lifetimes to restore clean water, food, and air. Everyone must get started with these cleaning processes. Every exposure to toxic water worsens the affects on our personal pollution burden and the disease associations or causes.

Disappearing rainforests, advancing deserts, melting ice caps, and glaciers, are large ongoing trends needing human attention and prompt action. Half of Earth's rainforests have been destroyed. More of the remaining rainforests are being lost every day. Most reasons for this are directly related to Earth's increasing population. Rainforests supply the Earth with water and oxygen. People need forests, oxygen, clean water, food, and rain to survive.

The arctic icecap will soon be totally melted sometime in the summer months. That hasn't happened in recorded history.[226] Vast portions of frozen "Greenland" are melting. Some sections are not re-freezing. Temporary progress may not reverse long term trends. Water can freeze and ice can return, but extinction of any life forms is permanent.

Nature isn't making many glaciers now. Glaciers are "calving" or breaking down and crashing into the seas. Glaciers calving sound like thunderclaps. That pseudo thunder sound is a real storm affecting the whole world. Freshwater disappears with the glaciers. Glacier National Park has few glaciers left. The remaining glaciers may be gone sometime in the foreseeable future before you get there. Glaciers are melting all over the planet. They will freeze and melt again. Glaciers have never melted from people's actions before. There is little snow left on Mount Kilimanjaro. The snow will be gone soon taking freshwater with it.

You valued what water meant before this book found you. Pick one water issue and work until it's fixed. Motivate everyone you know to work on and resolve just one water goal. The water you use daily should be the best you can get. That's the only way to make water better. If you don't act, today is the tomorrow you're worrying about. If today is the same as yesterday, you may have no tomorrow. There may be no tomorrow for children because there's no clean water for them today. The incidence of many childhood diseases rises every year.

Stop making profit and litter at Earth's expense. Earth's resources are limited. Make a good new working water system. Don't make the old bad water system bigger or faster. The current water system has its successes but is also failing at many levels. Parts of humanity fail with it.

People have filled the water with dangers like dichloro-diphenyl-trichloro-ethane or (DDT), polychlorinated biphenyls (PCBs), radiation, life threatening resistant bacteria, killer viruses, cancer-causing chemicals, plastics, mercury, lead, aluminum, other toxic metals, "artificial-foods, tastes, colors, and preservatives," and many other possible poisons harming and killing people and animals. Hurricanes, typhoons, and storms spread those toxins to areas thought safe before.

Reading that list won't satisfy your thirst. Is there no one stopping this craziness? The superfund was created in 1980. Funding for the superfund was stopped fifteen years later, although many old and new superfund sites are still harming people and other life forms. "Legacy" pollutants like PCBs, DDT, and others contaminate most superfund sites. Industries have been polluting many of those sites for more than fifty years. Most sites contain something that can kill you. Our solar system must recognize Earth as a superfund site.

The suffix "cide" means death. Pesticides, herbicides, insecticides, and other cides kill pests, plants, insects, people, and other life forms. All these "cides" have been found contaminating the water.

This planet contains water and land. Approximately 70 percent, plus or minus a couple percent, of planet Earth is covered with water of some form. The difficulty measuring produces the variety in estimates. We live on planet water. Water is survival. Canada has more freshwater than any other country. Thirty percent of Earth is land. Nine percent of the planet is desert. As we use too much of Earth's water, we expose more sand. Humans, not other life forms, overcrowd the land. People are the infestation. Good clean productive planets with water, food, and air supporting life forms are hard to find. Even if there is a planet B somewhere someday, and some humans can actually get to and live there, there is no "Planet C, D, or E."

Ocean water is saltwater. Sand surrounds the oceans. Below the water is more sand. Above the water is the polluted sky. People are mostly saltwater. We must drink freshwater to maintain our internal saltwater environment.

Healthy adult human females are about 68 percent, plus or minus several percent, water when healthy. Water content is a couple percentage points higher for healthy males we think. Don't drink saltwater. Saltwater causes dehydration and other problems. Drinking saltwater can end your life. Water and sand know the secrets of life and the land.

Big islands like North America, Africa, and Eurasia, have desert regions. Some desert names like Sahara, Mojave, Kalahari, Gobi, and Sonoran are familiar. The Atacama Desert in Chile is the driest place on Earth. There is no water there.[227]

Most of the Earth is covered with water. No job is more important than keeping water clean. To live well, water quality must be constantly improved and then maintained. Worsening water quality threatens the survival of life forms. We wouldn't have to clean water if we didn't dirty it. Water is the victim of people's indifference. Stop dirtying the water.

In 2014, the planet's water and land supported more than seven and less than eight billion people. Before 2050, the population is estimated to reach more than nine billion humans. About half of the occupied land has been made into cities. Currently more than half of Earth's population, four to five billion people, live in those cities. Clean non-toxic water supplies decrease as city populations increase, with only few small exceptions.

There is more bad news about water every day. People die from toxic water. The death of humans and other life forms increases as it follows water's downfall. Sorry, maybe it is being led downhill by unclean or contaminated water. Billions of people don't have clean water to drink. Still more billions of people only have toxic bad water to drink.

Many people drinking only bad water don't have enough bad drinking water either. This is the present now, not just the future.

It has been reported, the worst drought in recorded United States history occurred in 2012. It was worse than the drought that caused the Dust Bowl of the 1930's. It was worse than eighty years ago when forty-six of the forty-eight continental states experienced water shortages.[228] Scarce water supply is causing massive forest die-offs worldwide. Drought causes human die-offs as well. Fire and its suppression adds to the various prices paid in many forms.

People talk about managing water. Realize, and believe this, water manages people. If someone gets control of the water, they'll have power over people and other life forms.

CHAPTER 11

Fresh clean uncontaminated water, now unfindable outside anywhere should be wholly located everywhere for life to be and stay alive, survive, and thrive. The world's clean water begins at home. Polluted water falls through the sky when rain falls from anywhere to everywhere thanks to the clouds, wind, and mostly the people. Many rainforests are gone, going, or simply and openly contaminated. Some rainforests don't have rain falling every day like before. Simply and openly doesn't accurately describe the complicated, catastrophic issue worldwide pollution is and must be realized for changes to take place and improvements to occur. I don't need to hear or read more about pollution. I need to see actions benefiting the water and the environment locally and globally to believe there can be positive change now or future success later.

Keeping water clean is everyone's job. Get to it now. Time is wasting. Healthy life from clean water is disappearing. Sickness from dirty water is everywhere.

Air pollution contributes to water pollution. Air pollution worsens toxicity around the planet however much time it takes to travel, fly, or be carried by natural forces in air, water, on land, or by people. Weather patterns, human and animal migrations, insect flourishing, algae blooms, and other expected Earth activities force Earth changes whichever needs to be managed a priori. Either can be neither but never isn't ever. Rain falls into water already polluted from the sky, other "natural source" waters, and multiple toxic substances from people contamination origins. Some years and many miles ago, I saw a freshwater stream cascading down through boulders in Eastern Pennsylvania. Later, I saw a swift moving tidal saltwater creek emptying across the beach into the Delaware Bay near Cape May Point, New Jersey.

Soap bubbles floated near rocks, forgotten wood chunks, fallen trees, decaying stumps, and litter in both areas. Wading into the water, I grab handfuls of foam. The bubbles are gray and tan. They feel slippery and soapy. The smell of trash and sewage fills both surrounding air spaces and waters. Time makes the rotten smell as saturated in air as the fragrant sweet smell in springtime of flowers blooming and fruit ripening. Those odors cannot be avoided.

I work to provide information both deliberately and unintentionally to use, think, and create in many situations, whether in a fight, in flight, frozen in fright, or happily at peace. Pick your poison or antidote. When need be, keep your thinking unsullied by any incorrect thoughts of others. Written, spoken, and recorded words, imagined and real, unimaginable and computer-generated effects may affect people reading this. Read on at your own risk or pleasure. Whatever you rethink, evolve, and develop further, learn, or assimilate, make those thoughts integral to the way you behave.

Both water and sky were magical that night in New Jersey. A beautiful sunset happened with the sun drowning slowly into the boiling stale gray water of Delaware Bay. Daylight became darkness as the sunlight danced briefly with every ROY G BIV branch contemporary artist before winking out for the night.

Behind me, the beautiful moon sparkled, rising from the Atlantic Ocean. Moon drops floated into the sky. Colors exploded as moonlight kissed the fading sunlight racing the night. The bright day wound down as the brilliant moonlit starry sky exploded. The sunlight was drowning with the moon rising. Were they both trying to escape something they see we don't?

The slippery feel and foul smell of the suds confirmed and worsened my conclusions. Harmful looking or smelly water is dangerous. Remove any trash you see. Even water looking clean can be dirty. Most things in water, alive or toxic, trying to hurt or kill us, can't be seen without bringing you their danger. Be cautious and conservative about water quality. Prove water's safety before you believe or use it. Use distilled water whenever you can for whatever you need water for. It will expand the flavors you prefer, rather than contaminate your tastes with more toxins consumed.

On my way home, I went into a store selling "Water Ice." Their water ice is made from tap water, artificial flavor, artificial color, and is sweetened with white sugar made by genetic modification of sugar beets. The manufactured sugar is further sprayed with potassium bromate making it white like "real?" sugar. Not all water ice products are the same. Except for the bad water there's nothing "natural" in this aforementioned product. All the components have likely toxicities. I write "likely" toxicities to be fair when harm is suspected but not objectively proven. Many water toxicities are proven repeatedly. Out of multiple "possible" toxicities, one or more may be real. Some or all might be wrong. Maybe they're all true?

Chlorophyll makes plants green. Those plants cause no problems associated with excess chlorophyll consumption in most people. Chlorophyll is to plants as hemoglobin is to bloody life forms. Rarely chlorophyll is associated with minor itching and some stomach effects referred to as "side effects." As action chemicals go, think of dosage and speed to the expected effects, positive or negative. People mistakenly think if a little bit is good, more is better, and too much is just about enough. Whether a little bit, more, or too much, all amounts can also cause toxic effects. Anaphylaxis kills life forms.

There are no side effects of medicines. Everything medicine does to you is an effect, whether positive or negative, wanted or not.

Chlorophyll cleans your body. Chlorophyll has numerous theories about positive actions. In my opinion, some thoughts may be true yet unproven. More hypotheses may be "proven" to be untrue yet still useful according to current thinking and behavior. Some things not real can feel real. Disproving something everyone thinks is true is difficult to do. Water is no exception. *Sine que non*. Prove it. Find proof where you can.

The current water paradigm must change. In my opinion, no price is too high to pay for clean non-toxic water to help people, and other life-forms, be and stay healthy. Being unaware about water toxicity may make you sick or kill you. Some water-borne rare infections can end your life in less than a day. Hospitals and health systems keep records of water borne illnesses. Cancer often takes longer to end your life. Toxins in water can cause both, either, or neither. All are dependent on the person's underlying general health or present diseases. Poor water quality should not be tolerated, ignored, or proceed unfixed.

You can't just improve a water system failing everywhere. Failure in water anywhere threatens water safety and health for life-forms everywhere. Failure needs revolution not evolution. Evolution takes time we may not have. The cost is rising. Literally, good clean water is now bought and sold on a worldwide financial market exchange. What will money, profit, and cost do to the current water system? Replace the flawed water plan with a working water structure. People need water providing value, not toxicity or harm. Water should make positive change for all life forms. Water must not make you sick and we must do what people need to stay healthy. Before the dream of health is realized, we must end the nightmare of water pollution. Nightmares are dead dreams.

Read Water: Warnings and Rewards while there's still time to fix water. Fix your personal water, organs, physiologic water-systems, any pathological diseases, and the rest of your body parts first. To survive, fix all water around you. Get started and stay working. This planet's water needs the greatest, worldwide effort by everyone, everywhere, immediately to have a chance at a world water fix. It's later than we think.

The world has four times more people in the early twenty-first century than eighty years earlier at the end of World War II. The population will soon number over nine billion bags of saltwater disguised as Earth humans. Freshwater needs increase as clean water supplies decrease. That gap sounds an alarm signaling danger. Fewer people on the planet would use less water for basic human survival. More people currently alive would survive if fewer people died from problems with water. If people don't act, the population will reduce itself through the water it has had for hundreds of years before.

Too much or too little water on the planet, or in the wrong places while not in the right places, have always been emergent problems. Those emergencies often result in human diseases, death, and destruction.

Twenty-five years into this century most teenagers of today won't yet be thirty years old. How must it feel to face the problems of not enough freshwater, drinking water, or not enough bad water, when your entire cohort, your tribe, or generation was under thirty years old?

We must face water problems now. The needed "now" changes would have been more effective if done "Before." The quality of products depends on the purity of water used to make or grow them. Most things you buy to drink are made from bad water. Waters claiming to be good are made from bad water. Numerous toxins aren't removed from water unless the water is steam distilled. All real distillation requires steam produced by heat. Be careful of other "pseudo-distillation" terms.

After reading the title of this book you know how I feel about water products. What are the consequences of consuming water toxins? Fatal cancer is one awful outcome. Thank you for reading this book. You're the people who know about the problems with water. You're the folks who want to know more and make positive permanent change. You're the people who will make a difference. We must become the targeted change we need to see; changes humanity must observe and fix to survive. The best time to make better water was years and more years ago. The next best time to fix the water is today.

The end, but only for now.

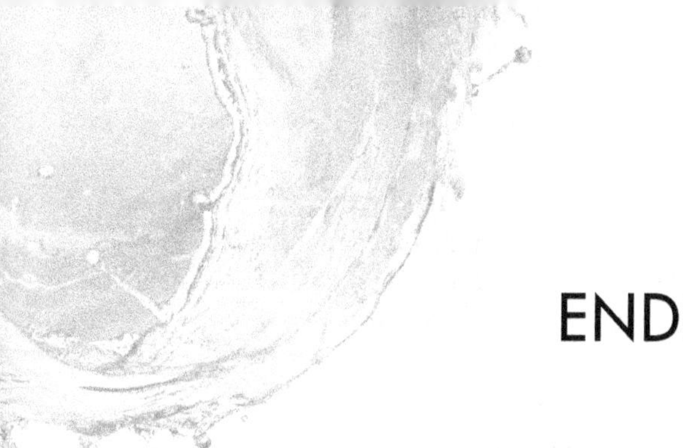

ENDNOTES

1 Sbichao. "Cancer Rates in Some NJ Counties Are Very High." New Jersey 101.5 – Proud to Be New Jersey – New Jersey News Radio, 28 Oct. 2017, nj1015.com/new-jerseys-high-cancer-rates-how-your-county-measures-up/.

2 "Cancer Prevention and Control." *Centers for Disease Control and Prevention*, Centers for Disease Control and Prevention, 2 June 2017, www.cdc.gov/cancer/dcpc/data/state.htm.

3 "History of the Clean Water Act." *EPA*, Environmental Protection Agency, 8 Aug. 2017, www.epa.gov/laws-regulations/history-clean-water-act.

4 "Extreme Philly Fishing." *Are the Fish in the Schuylkill Edible? Pollutants: Myth or Reality?*, extremephillyfishing.blogspot.com/2011/11/are-fish-in-schuylkill-edible.html.

5 Dattaro, Laura. "The Most Widely Used Herbicide in the United States Could Cause Cancer in Humans, Says a World Health Organization Study." *VICE News*, 23 Mar. 2015, news.vice.com/article/the-most-widely-used-herbicide-in-the-united-states-could-cause-cancer-in-humans-says-a-world-health-organization-study.

6 *Labels & MSDS | Roundup*, www.roundup.ca/en/labels-msds.

7 User, Super. "Home." *Bio Minerals Technologies*, www.biomineralstechnologies.com/farm-solutions/reduce-chemicals/effects-of-glyphosate-on-soils-and-plants.

8 *Labels & MSDS | Roundup*, www.roundup.ca/en/labels-msds.

[9] Gomes, et al. "Alteration of Plant Physiology by Glyphosate and Its by-Product Aminomethylphosphonic Acid: an Overview | Journal of Experimental Botany | Oxford Academic." *OUP Academic*, Oxford University Press, 19 July 2014, academic.oup.com/jxb/article/65/17/4691/558768.

[10] User, Super. "Home." *Bio Minerals Technologies*, www.biomineralstechnologies.com/farm-solutions/reduce-chemicals/effects-of-glyphosate-on-soils-and-plants.

[11] Stone, Judy. "Antibiotic Resistance From Unexpected Sources—Herbicides, Dust And Metals." *Forbes*, Forbes Magazine, 1 Apr. 2015, www.forbes.com/sites/judystone/2015/04/01/antibiotic-resistance-from-unexpected-sources/#1e2d5f8bdbca.

[12] User, Super. "Home." *Bio Minerals Technologies*, www.biomineralstechnologies.com/farm-solutions/reduce-chemicals/effects-of-glyphosate-on-soils-and-plants.

[13] Stone, Judy. "Antibiotic Resistance From Unexpected Sources—Herbicides, Dust And Metals." *Forbes*, Forbes Magazine, 1 Apr. 2015, www.forbes.com/sites/judystone/2015/04/01/antibiotic-resistance-from-unexpected-sources/#1e2d5f8bdbca.

[14] Battaglin, William A. *Glyphosate Herbicide Found in Many Midwestern Streams, Antibiotics Not Common*, toxics.usgs.gov/highlights/glyphosate02.html.

[15] Majewski, M S, et al. "Pesticides in Mississippi Air and Rain: a Comparison between 1995 and 2007." *Environmental Toxicology and Chemistry.*, U.S. National Library of Medicine, June 2014, www.ncbi.nlm.nih.gov/pubmed/24549493.

[16] http://www.ithaka-journal.net/herbizide-im-urine.

[17] Stone, Judy. "Antibiotic Resistance From Unexpected Sources—Herbicides, Dust And Metals." *Forbes*, Forbes Magazine, 1 Apr. 2015, www.forbes.com/sites/judystone/2015/04/01/antibiotic-resistance-from-unexpected-sources/#1e2d5f8bdbca.

[18] "Glyphosate Use Worldwide 1994-2014 | Statistic." *Statista*, www.statista.com/statistics/567250/glyphosate-use-worldwide/.

[19] ecowatch_contributor. "15 Health Problems Linked to Monsanto's Roundup." *EcoWatch*, EcoWatch, 27 June 2016, www.ecowatch.com/15-health-problems-linked-to-monsantos-roundup-1882002128.html.

[20] https://www.iarc.fr/en/media-centre/iarcnews/pdf/MonographVolume112.pdf.

[21] "EPA Releases Draft Risk Assessments for Glyphosate." *EPA*, Environmental Protection Agency, 18 Dec. 2017.

[22] Agricultural Health Study." *National Institutes of Health*, U.S. Department of Health and Human Services, aghealth.nih.gov/.

[23] "Roundup Lawsuit." *ConsumerSafety.org*, www.consumersafety.org/legal/roundup-lawsuit/.

[24] APPDMZ\ccvivr_monsanto. "Products." *Monsanto* | 한 눈에 보는 몬산토, www.monsantoglobal.com/global/au/products/Pages/roundup-ready-flex.aspx.

[25] "What Does Genetically Modified (GMO) Mean?" *EatingWell*, EatingWell, www.eatingwell.com/video/6923/what-does-genetically-modified-mean/.

[26] "Roundup Ready Crops." *Jon*, web.mit.edu/demoscience/Monsanto/about.html.

[27] Eenennaam, Alison L Van. *Advances in Pediatrics.*, U.S. National Library of Medicine, 2013, www.ncbi.nlm.nih.gov/pmc/articles/PMC4015968/.

[28] Blue, John. "Dr. Alison Van Eenennaam–The GMO Debate: Twenty Years of Animal Hea..." *LinkedIn SlideShare*, 22 Dec. 2016, www.slideshare.net/trufflemedia/dr-alison-van-eenennaam-the-gmo-debate-twenty-years-of-animal-health-and-livestock-feeding-studies.

[29] http://agpolicy.org/weekpdf/202.pdf.

[30] Ibid.

[31] Kershner, Kate. "How Reverse Osmosis Works." *HowStuffWorks Science*, HowStuffWorks, 8 Mar. 2018, science.howstuffworks.com/reverse-osmosis2.htm.

[32] US Department of Veterans Affairs, and Veterans Health Administration. "Public Health." *World War II Exposures–Public Health*, 25 Nov. 2013, www.publichealth.va.gov/exposures/agentorange/basics.asp.

[33] APPDMZ\ccvivr_monsanto. "Products." *Monsanto* | 한 눈에 보는 몬산토, www.monsantoglobal.com/global/au/products/Pages/roundup-ready-flex.aspx.

[34] US Department of Veterans Affairs, and Veterans Health Administration. "Public Health." *World War II Exposures–Public Health*, 25 Nov. 2013, www.publichealth.va.gov/exposures/agentorange/basics.asp.

[35] "Roundup Lawsuit." *ConsumerSafety.org*, www.consumersafety.org/legal/roundup-lawsuit/.

[36] US Department of Veterans Affairs, and Veterans Health Administration. "Public Health." *World War II Exposures–Public Health*, 25 Nov. 2013, www.publichealth.va.gov/exposures/agentorange/basics.asp.

[37] Ibid.

[38] "Vietnam's Horrific Legacy: The Children of Agent Orange." *NewsComAu*, 25 May 2015, www.news.com.au/world/asia/vietnams-horrific-legacy-the-children-of-agent-orange/news-story/c008ff36ee3e840b005405a55e21a3e1.

[39] History.com Staff. "Agent Orange." *History.com*, A&E Television Networks, 2011, www.history.com/topics/vietnam-war/agent-orange.

[40] "Roundup Ready Crops." *Jon*, web.mit.edu/demoscience/Monsanto/about.html.

[41] History.com Staff. "Agent Orange." *History.com*, A&E Television Networks, 2011, www.history.com/topics/vietnam-war/agent-orange.

[42] "Learn about Dioxin." *EPA*, Environmental Protection Agency, 22 Mar. 2018, www.epa.gov/dioxin/learn-about-dioxin.

[43] "A Quote by Sherry A. Rogers." *Goodreads*, Goodreads, www.goodreads.com/quotes/928715-the-solution-to-pollution-is-dilution-it-is-very-logical.

[44] Rafferty, John P. "Just How Many Oceans Are There?" *Encyclopædia Britannica*, Encyclopædia Britannica, Inc., www.britannica.com/story/just-how-many-oceans-are-there.

[45] Aktar, Md. Wasim, et al. *Interdisciplinary Toxicology*, Slovak Toxicology Society SETOX, Mar. 2009, www.ncbi.nlm.nih.gov/pmc/articles/PMC2984095/.

[46] Ibid.

[47] "Numbers of Insects (Species and Individuals)." *Smithsonian Institution*, www.si.edu/spotlight/buginfo/bugnos.

[48] Alavanja, Michael C.R. *Reviews on Environmental Health*, U.S. National Library of Medicine, 2009, www.ncbi.nlm.nih.gov/pmc/articles/PMC2946087/.

[49] Aktar, Md. Wasim, et al. *Interdisciplinary Toxicology*, Slovak Toxicology Society SETOX, Mar. 2009, www.ncbi.nlm.nih.gov/pmc/articles/PMC2984095/.

[50] Perlman, Howard, and USGS. "Pesticides in Groundwater." *Pesticides in Groundwater*, water.usgs.gov/edu/pesticidesgw.html.

[51] Ibid.

[52] "About Neonicotinoids." *Pesticide Action Network UK*, www.pan-uk.org/about neonicotinoids/.

[53] "Why Europe's Insecticide Ban Is Big News for Bees." *National Geographic*, National Geographic Society, 27 Apr. 2018, news.nationalgeographic.com/2018/04/neonics-neonicotinoids-banned-european-union-protect-bees-pollinators-environment-science-spd/.

[54] *ASHS*, www.ashs.org/blogpost/1288786/251171/U-S-States-Begin-Ban-on-Neonicotinoids.

[55] Guarino, Ben. "First Evidence Found of Popular Farm Pesticides in Drinking Water." *The Washington Post*, WP Company, 5 Apr. 2017, www.washingtonpost.com/news/speaking-of-science/wp/2017/04/05/iowa-scientists-find-first-evidence-of-popular-farm-pesticides-in-drinking-water/?noredirect=on&utm_term=.3167ac21f24d.

[56] "Occurrence of Neonicotinoid Insecticides in Finished Drinking Water and Fate during Drinking Water Treatment." *ACS Publications*, pubs.acs.org/doi/abs/10.1021/acs.estlett.7b00081.

[57] Environmental Working Group. "Dirty Dozen™ Fruits and Vegetables with the Most Pesticides." *EWG Tap Water Database*, www.ewg.org/foodnews/dirty-dozen.php.

[58] FRIIS-HANSEN, BENT J., et al. "TOTAL BODY WATER IN CHILDREN." *Pediatrics*, American Academy of Pediatrics, 1 Mar. 1951, pediatrics.aappublications.org/content/7/3/321.

[59] Perlman, Howard, and USGS. "The Water in You." *Livestock Water Use, the USGS Water Science School*, water.usgs.gov/edu/propertyyou.html.

[60] "Dehydration." *Mayo Clinic*, Mayo Foundation for Medical Education and Research, 15 Feb. 2018, www.mayoclinic.org/diseases-conditions/dehydration/symptoms-causes/syc-20354086.

[61] "How Much Salt Is in a Human Body?" *Science Focus*, www.sciencefocus.com/qa/how-much-salt-human-body.

[62] "Dehydration and Aging." *Senior Immunizations*, www.johnmuirhealth.com/health-education/health-wellness/senior_health/dehydration-aging.html.

[63] Batmanghelidj, F., Your Body's Many Cries For Water, You aren't sick you're thirsty. Global Health Solutions, Vienna, Virginia, 1992.

[64] "How Much Salt Is in a Human Body?" *Science Focus*, www.sciencefocus.com/qa/how-much-salt-human-body.

65 Batmanghelidj, F., Your Body's Many Cries For Water, You aren't sick you're thirsty. Global Health Solutions, Vienna, Virginia, 1992.

66 Ibid.

67 Goldbaum, Kate. "Does Caffeine Really Dehydrate You?" *LiveScience*, Purch, 21 July 2016, www.livescience.com/55479-does-caffeine-cause-dehydration.html.

68 "Decaf Coffee Isn't Actually Caffeine-Free–Here's How Much Caffeine You're Really Drinking." *Philippines to Shut Polluted Isle Duterte Called a Cesspool*, www.msn.com/en-us/foodanddrink/nonalcoholic/decaf-coffee-isnt-actually-caffeine-free-—--heres-how-much-caffeine-youre-really-drinking/ar-BBEMtBb.

69 Glazer, James L. "Management of Heatstroke and Heat Exhaustion." *American Family Physician*, 1 June 2005, www.aafp.org/afp/2005/0601/p2133.html.

70 Ibid.

71 Ibid.

72 "Dehydration: How Lack of Water Affects Your Body." *MSC Nutrition*, 7 Mar. 2017, www.msc-nutrition.co.uk/ahaide/.

73 "Drinking Water Contaminants - Standards and Regulations." *EPA*, Environmental Protection Agency, 22 May 2017, www.epa.gov/dwstandardsregulations.

74 French, Andrew. "Difference Between Distilled Water & Drinking Water." *Healthy Eating | SF Gate*, 11 June 2018, healthyeating.sfgate.com/difference-between-distilled-water-drinking-water-9002.html.

75 Shaw, Gina. "Water and Your Diet: Staying Slim and Regular With H2O." *WebMD*, WebMD, www.webmd.com/diet/features/water-for-weight-loss-diet#1.

76 Burn, Life by Daily. "What Are Natural Flavors, Really?" *CNN*, Cable News Network, 14 Jan. 2015,

www.cnn.com/2015/01/14/health/feat-natural-flavors-explained/index.html.

77 Young, Robert O., Young, Shelley Redford, *The pH Miracle, Balance Your Diet, Reclaim Your Health,* Warner Books, New York, 2002.

78 Kershner, Kate. "How Reverse Osmosis Works." *HowStuffWorks Science*, HowStuffWorks, 8 Mar. 2018, science.howstuffworks.com/reverse-osmosis2.htm.

79 Contributors, HowStuffWorks.com. "How to Make Distilled Water." *HowStuffWorks*, HowStuffWorks, 4 Apr. 2011, home.howstuffworks.com/green-living/how-to-make-distilled-water.htm.

80 "Natural Health Information Articles and Health Newsletter by Dr. Joseph Mercola." *Mercola.com*, British Medical Journal, www.mercola.com/article/water/distilled_water.htm.

81 Admingen. "Is Distilled Water More Dangerous Than Tap Water?" *Health News from Hallelujah Diet*, 1 Oct. 2014, www.myhdiet.com/healthnews/rev-malkmus/is-distilled-water-more-dangerous-than-tap-water/.

82 "Mineral Deficiency | Definition and Patient Education." *Healthline*, Healthline Media, www.healthline.com/health/mineral-deficiency#1.

83 https://www.cdc.gov/healthywater/drinking/public/drinking-water-faq.html.

84 Ibid.

85 Ibid.

86 "Bottled Water vs. Filtered Tap Water | Care2 Healthy Living." *Care2 Causes*, www.care2.com/greenliving/bottled-water-vs-filtered-tap-water.html.

87 Burn, Life by Daily. "What Are Natural Flavors, Really?" *CNN*, Cable News Network, 14 Jan. 2015, www.cnn.com/2015/01/14/health/feat-natural-flavors-explained/index.html.

[88] Glazer, James L. "Management of Heatstroke and Heat Exhaustion." *American Family Physician*, 1 June 2005, www.aafp.org/afp/2005/0601/p2133.html.

[89] "The Five Steps of the Scientific Method." *Actforlibraries.org*, www.actforlibraries.org/the-five-steps-of-the-scientific-method/.

[90] Lerner, Ab. "Top 30 Hydrating Foods." *Shape Magazine*, Shape Magazine, 10 Dec. 2015, www.shape.com/healthy-eating/healthy-drinks/top-30-hydrating-foods.

[91] Ibid.

[92] Carson, Rachel. *Silent Spring*. *AbeBooks*, Greenhaven Pr, 1 Jan. 1970, www.abebooks.com/9780618249060/Silent-Spring-Rachel-Carson-0618249060/plp.

[93] Cabral, João P. S. *Advances in Pediatrics.*, U.S. National Library of Medicine, Oct. 2010, www.ncbi.nlm.nih.gov/pmc/articles/PMC2996186/.

[94] Ibid.

[95] "Drink 8 Glasses of Water a Day: Fact or Fiction?" *Healthline*, Healthline Media, www.healthline.com/nutrition/8-glasses-of-water-per-day.

[96] Huff, James, and Peter F. Infante. *Mutagenesis*, Oxford University Press, Sept. 2011, www.ncbi.nlm.nih.gov/pmc/articles/PMC3165940/.

[97] Carson, Rachel. *Silent Spring*. *AbeBooks*, Greenhaven Pr, 1 Jan. 1970, www.abebooks.com/9780618249060/Silent-Spring-Rachel-Carson-0618249060/plp.

[98] Huff, James, and Peter F. Infante. *Mutagenesis*, Oxford University Press, Sept. 2011, www.ncbi.nlm.nih.gov/pmc/articles/PMC3165940/.

[99] Ibid.

[100] Fredericks, Bob. "City's Styrofoam Ban Goes into Effect Jan. 1." *New York Post*, New York

Post, 14 June 2018, nypost.com/2018/06/13/citys-styrofoam-ban-goes-into-effect-january-1/.

[101] "6 Sneaky Cancer Culprits." *ABC News*, ABC News Network, abcnews.go.com/Health/Wellness/sneaky-cancer-culprits/story?id=19475248.

[102] Huff, James, and Peter F. Infante. *Mutagenesis*, Oxford University Press, Sept. 2011, www.ncbi.nlm.nih.gov/pmc/articles/PMC3165940/.

[103] Center for Food Safety and Applied Nutrition. "Public Health Focus–Bisphenol A (BPA): Use in Food Contact Application." *U S Food and Drug Administration Home Page*, Center for Food Safety and Applied Nutrition, www.fda.gov/newsevents/publichealthfocus/ucm064437.htm.

[104] Huff, James, and Peter F. Infante. *Mutagenesis*, Oxford University Press, Sept. 2011, www.ncbi.nlm.nih.gov/pmc/articles/PMC3165940/.

[105] Center for Food Safety and Applied Nutrition. "Public Health Focus–Bisphenol A (BPA): Use in Food Contact Application." *U S Food and Drug Administration Home Page*, Center for Food Safety and Applied Nutrition, www.fda.gov/newsevents/publichealthfocus/ucm064437.htm.

[106] Ibid.

[107] "Bisphenol A (BPA)." *National Institute of Environmental Health Sciences*, U.S. Department of Health and Human Services, www.niehs.nih.gov/health/topics/agents/sya-bpa/index.cfm.

[108] Ibid.

[109] Ibid.

[110] "Toxic Chemicals Found in Minority Cord Blood." *EWG*, www.ewg.org/news/news-releases/2009/12/02/toxic-chemicals-found-minority-cord-blood.

[111] https://www.niehs.nih.gov/health/materials/endocrine_disruptors_508.pdf.

[112] Ibid.

[113] "Exposure to Chemicals in Plastic." *Breastcancer.org*, www.breastcancer.org/risk/factors/plastic.

[114] "Environmental Factor–April 2014: New Study Links BPA and Prostate Cancer in Humans." *National Institute of Environmental Health Sciences*, U.S. Department of Health and Human Services, factor.niehs.nih.gov/2014/4/science-bpaprostate/index.htm.

[115] Tarapore P, Ying J, Ouyang B, Burke B, Bracken B, and Ho SM. 2014. Exposure to bisphenol A correlates with early-onset prostate cancer and promotes centrosome amplification and anchorage-independent growth in vitro. PLoS One 9(3):e90332.

[116] "Environmental Factor–April 2014: New Study Links BPA and Prostate Cancer in Humans." *National Institute of Environmental Health Sciences*, U.S. Department of Health and Human Services, factor.niehs.nih.gov/2014/4/science-bpaprostate/index.htm.

[117] Tarapore P, Ying J, Ouyang B, Burke B, Bracken B, and Ho SM. 2014. Exposure to bisphenol A correlates with early-onset prostate cancer and promotes centrosome amplification and anchorage-independent growth in vitro. PLoS One 9(3):e90332.

[118] Ibid.

[119] "Exposure to Chemicals in Plastic." *Breastcancer.org*, www.breastcancer.org/risk/factors/plastic.

[120] Leonardi, Alberto, et al. *International Journal of Environmental Research and Public Health*, MDPI, Sept. 2017, www.ncbi.nlm.nih.gov/pmc/articles/PMC5615581/.

[121] Provvisiero, Donatella Paola, et al. *International Journal of Environmental Research and Public Health*, MDPI, Oct. 2016, www.ncbi.nlm.nih.gov/pmc/articles/PMC5086728/.

[122] T., Do Minh, et al. *Health Promotion and Chronic Disease Prevention in Canada : Research, Policy and Practice*, Public

Health Agency of Canada, Dec. 2017, www.ncbi.nlm.nih.gov/pmc/articles/PMC5765817/.

123 Roen, Emily L., et al. *Environmental Research*, U.S. National Library of Medicine, Oct. 2015, www.ncbi.nlm.nih.gov/pmc/articles/PMC4545741/.

124 Provvisiero, Donatella Paola, et al. *International Journal of Environmental Research and Public Health*, MDPI, Oct. 2016, www.ncbi.nlm.nih.gov/pmc/articles/PMC5086728/.

125 Bilbrey, Jenna. "BPA-Free Plastic Containers May Be Just as Hazardous." *Scientific American*, 11 Aug. 2014, www.scientificamerican.com/article/bpa-free-plastic-containers-may-be-just-as-hazardous/.

126 "Bisphenol A (BPA)." *National Institute of Environmental Health Sciences*, U.S. Department of Health and Human Services, www.niehs.nih.gov/health/topics/agents/sya-bpa/index.cfm.

127 "Exposure to Chemicals in Plastic." *Breastcancer.org*, www.breastcancer.org/risk/factors/plastic.

128 https://www.greensheepwater.com/plastic-problems/.

129 Ehlers, Lance. "The Graduate 'One Word: Plastics.'" *YouTube*, YouTube, 9 Nov. 2007, www.youtube.com/watch?v=PSxihhBzCjk.

130 "A Whopping 91% of Plastic Isn't Recycled." *National Geographic*, National Geographic Society, 16 May 2018, news.nationalgeographic.com/2017/07/plastic-produced-recycling-waste-ocean-trash-debris-environment/.

131 "Bottled Water Brands | Nestlé Waters North America." *Https://Www.nestle-Watersna.com*, www.nestle-watersna.com/en/bottled-waterbrands?&iq_id=105709464VQ16c&utm_medium=cpc&utm_source=google&utm_campaign=Corporate_PR_Who_We_Are_Brand_MBR&utm_term=105709464-VQ16-c&ds_kid=43700031108710589&gclid=Cj0KCQiAr93gBRDSARIsADvHiOpNGNroY7Rn9_o4ghobyhuUB3s3VIXoeFujdLkH9HA6ZtzYL1vWgscaAjeoEALws_wcB&gclsrc=aw.ds.

[132] "Piling up: Drowning in a Sea of Plastic." *CBS News*, CBS Interactive, 5 Aug. 2018, www.cbsnews.com/news/piling-up-drowning-in-a-sea-of-plastic/.

[133] Ibid.

[134] http://advances.sciencemag.org/content/3/7/e1700782.

[135] https://www.pbs.org/newshour/science/humans-made-8-3-billion-tons-plastic-go.

[136] Ibid.

[137] Ibid.

[138] Ibid.

[139] Lapidos, Juliet. "Do Plastic Bags Really Take 500 Years to Break down in a Landfill?" *Slate Magazine*, Slate, 27 June 2007, slate.com/news-and-politics/2007/06/do-plastic-bags-really-take-500-years-to-break-down-in-a-landfill.html.

[140] The Collins Law Firm, P.C. "How Landfills Contaminate the Groundwater." *The Collins Law Firm, P.C.*, The Collins Law Firm, P.C., 11 Apr. 2018, www.collinslaw.com/blog/2017/05/how-landfills-contaminate-the-groundwater.shtml.

[141] "A Whopping 91% of Plastic Isn't Recycled." *National Geographic*, National Geographic Society, 16 May 2018, news.nationalgeographic.com/2017/07/plastic-produced-recycling-waste-ocean-trash-debris-environment/.

[142] Ibid.

[143] Ibid.

[144] "PLASTIC CHINA Documentary Film Official Site." *PLASTIC CHINA Documentary Film Official Site*, www.plasticchina.org/.

[145] https://www.scmp.com/magazines/post-magazine/arts-music/article/2146688/plastic-china-documentary-focuses-human-impact.

[146] http://advances.sciencemag.org/content/3/7/e1700782.

[147] "A Whopping 91% of Plastic Isn't Recycled." *National Geographic*, National Geographic Society, 16 May 2018,

news.nationalgeographic.com/2017/07/plastic-produced-recycling-waste-ocean-trash-debris-environment/.

[148] http://advances.sciencemag.org/content/3/7/e1700782.

[149] "A Whopping 91% of Plastic Isn't Recycled." *National Geographic*, National Geographic Society, 16 May 2018, news.nationalgeographic.com/2017/07/plastic-produced-recycling-waste-ocean-trash-debris-environment/.

[150] "Eight Million Tons of Plastic Dumped in Ocean Every Year." National Geographic, National Geographic Society, 10 Oct. 2017, news.nationalgeographic.com/news/2015/02/150212-ocean-debris-plastic-garbage-patches-science/.

[151] "Piling up: Drowning in a Sea of Plastic." *CBS News*, CBS Interactive, 5 Aug. 2018, www.cbsnews.com/news/piling-up-drowning-in-a-sea-of-plastic/.

[152] https://www.pbs.org/newshour/science/humans-made-8-3-billion-tons-plastic-go.

[153] https://earthhow.com/ocean-currents/.

[154] https://www.cbsnews.com/news/great-pacific-gargage-patch-pacific-ocean-growing-rapidly-study-reveals-pollution/.

[155] Ibid.

[156] https://www.cbsnews.com/news/piling-up-drowning-in-a-sea-of-plastic/.

[157] Ibid.

[158] https://www.cbsnews.com/news/could-an-enzyme-eradicate-plastic-pollution-in-the-worlds-oceans/.

[159] "A Whopping 91% of Plastic Isn't Recycled." *National Geographic*, National Geographic Society, 16 May 2018, news.nationalgeographic.com/2017/07/plastic-produced-recycling-waste-ocean-trash-debris-environment/.

[160] https://www.cbsnews.com/news/could-an-enzyme-eradicate-plastic-pollution-in-the-worlds-oceans/.

[161] Ibid.

[162] "This Bug Can Eat Plastic. But Can It Clean Up Our Mess?" *National Geographic*, National Geographic Society, 24 Apr. 2017, news.nationalgeographic.com/2017/04/wax-worms-eat-plastic-polyethylene-trash-pollution-cleanup/.

[163] "Cleaning up the Plastic in the Ocean." *CBS News*, CBS Interactive, www.cbsnews.com/news/the-great-pacific-garbage-patch-cleaning-up-the-plastic-in-the-ocean-60-minutes/.

[164] CBS/AP. "Huge Floating Device Isn't Trapping Plastic in Pacific Ocean." *CBS News*, CBS Interactive, 18 Dec. 2018, www.cbsnews.com/news/great-pacific-garbage-patch-huge-floating-device-isnt-trapping-plastic-waste-in-pacific-ocean/.

[165] "Cleaning up the Plastic in the Ocean." *CBS News*, CBS Interactive, www.cbsnews.com/news/the-great-pacific-garbage-patch-cleaning-up-the-plastic-in-the-ocean-60-minutes/.

[166] CBS/AP. "Huge Floating Device Isn't Trapping Plastic in Pacific Ocean." *CBS News*, CBS Interactive, 18 Dec. 2018, www.cbsnews.com/news/great-pacific-garbage-patch-huge-floating-device-isnt-trapping-plastic-waste-in-pacific-ocean/.

[167] Ibid.

[168] Ibid.

[169] Ibid.

[170] Ibid.

[171] http://www.bagitmovie.com/.

[172] https://topdocumentaryfilms.com/addicted-plastic/.

[173] https://video.vice.com/en_us/video/garbage-island/563b9c912aab5c416bc75039.

[174] CBS/AP. "Huge Floating Device Isn't Trapping Plastic in Pacific Ocean." *CBS News*, CBS Interactive, 18 Dec. 2018, www.cbsnews.com/news/great-pacific-garbage-patch-huge-floating-device-isnt-trapping-plastic-waste-in-pacific-ocean/.

[175] Sumner, Thomas. "Arctic Ice Travels Fast, Carrying Pollution." *Science News for Students*, 19 Oct. 2016, www.sciencenewsforstudents.org/article/arctic-ice-travels-fast-carrying-pollution.

[176] "New Jersey Oil & Gas Fracking Issues Map," *Texas Map of Oil & Gas Wells*, www.drillingmaps.com/New-Jersey.html#.XDjrQ89KjBI.

[177] Rubright, Sam. "34 States in U.S. Have Active Oil & Gas Activity–Based on 2016 Analysis." *FracTracker Alliance*, Sam Rubright, DrPH Https://Www.fractracker.org/a5ej20sjfwe/Wp-Content/Uploads/2016/05/New-FT-Website-Logo.png, 23 Mar. 2017, www.fractracker.org/2017/03/34-states-active-drilling-2016/.

[178] "Three States Have Banned Fracking Practice." *Heal Naturally*, Heal Naturally, 7 Mar. 2018, www.realnatural.org/three-states-ban-fracking/.

[179] "What Is Fracking and Why Is It Controversial?" *BBC News*, BBC, 15 Oct. 2018, www.bbc.com/news/uk-14432401.

[180] *Chemservice.com.* www.chemservice.com/news/2014/06/chemicals-and-fracking-what-makes-up-fracking-fluid/.

[181] Ibid.

[182] "Gasland 2: A Film by Josh Fox–NOW on HBO." *Gasland*, www.gaslandthemovie.com/whats-fracking/faq/fracking-fluid.

[183] "Chemical Use In Hydraulic Fracturing." *FracFocus: Chemical Disclosure Registry*, fracfocus.org/water-protection/drilling-usage.

[184] "The Slickwater Story @NAShaleMag." *North American Shale Magazine*, northamericanshalemagazine.com/articles/711/the-slickwater-story.

[185] "How Much Water Does the Typical Hydraulically Fractured Well Require?" *American Geosciences Institute*, 15 Aug. 2017, www.americangeosciences.org/critical-issues/faq/how-much-water-does-typical-hydraulically-fractured-well-require.

186 Mulliniks, Brent. "Recycling Fracking Wastewater." *Valve Magazine*, www.valvemagazine.com/magazine/sections/water-works/5579-recycling-fracking-wastewater.html.

187 "Fracking Wastewater Management." *WaterWorld*, 1 Nov. 2013, www.waterworld.com/articles/wwi/print/volume-28/issue-5/regional-spotlight-us-caribbean/fracking-wastewater-management.html.

188 Mulliniks, Brent. "Recycling Fracking Wastewater." *Valve Magazine*, www.valvemagazine.com/magazine/sections/water-works/5579-recycling-fracking-wastewater.html.

189 Ibid.

190 *Exxon Valdez Oil Spill*, response.restoration.noaa.gov/about/media/train-derails-paulsboro-nj-releasing-23000-gallons-toxic-vinyl-chloride-gas.html.

191 "Toxic Substances Portal–Vinyl Chloride." *Centers for Disease Control and Prevention*, Centers for Disease Control and Prevention, 14 Aug. 2018, www.atsdr.cdc.gov/ToxProfiles/TP.asp?id=282&tid=51.

192 REFERENCE "Vinyl Chloride." *National Cancer Institute*, www.cancer.gov/about-cancer/causes-prevention/risk/substances/vinyl-chloride.

193 https://www.popularmechanics.com/technology/gadgets/a1327/4212536/.

194 *Exxon Valdez Oil Spill*, response.restoration.noaa.gov/about/media/train-derails-paulsboro-nj-releasing-23000-gallons-toxic-vinyl-chloride-gas.html.

195 http://www.delawareriverkeeper.org/sites/default/files/A_RIVER_AGAIN_2012.pdf.

196 "DELAWARE CRASH CREATES OIL SPILL." *The New York Times*, The New York Times, 20 Feb. 1974, www.nytimes.com/1974/02/20/archives/delaware-crash-creates-oil-spill-tanker-in-trouble-st-side-of.html.

197 Rotman, Michael. "Cuyahoga River Fire." *Cleveland Historical*, clevelandhistorical.org/items/show/63.

[198] "Sewage Treatment–Septic Tanks." *Bacteria, Microorganisms, Drinking, and Microbes–JRank Articles*, science.jrank.org/pages/6089/Sewage-Treatment-Septic-tanks.html.

[199] Schwirtz, Michael. "Report Cites Large Release of Sewage From Hurricane Sandy." *The New York Times*, The New York Times, 19 Oct. 2018, www.nytimes.com/2013/05/01/nyregion/hurricane-sandy-sent-billions-of-gallons-of-sewage-into-waterways.html.

[200] Specter, Michael. "Ocean Dumping Is Ending, but Not Problems; New York Can't Ship, Bury or Burn Its Sludge, but No One Wants a Processing Plant." *The New York Times*, The New York Times, 29 June 1992, www.nytimes.com/1992/06/29/nyregion/ocean-dumping-ending-but-not-problems-new-york-can-t-ship-bury-burn-its-sludge.html.

[201] "Potential Well Water Contaminants and Their Impacts." *EPA*, Environmental Protection Agency, 23 Feb. 2018, www.epa.gov/privatewells/potential-well-water-contaminants-and-their-impacts.

[202] Ibid.

[203] Ibid.

[204] https://dhss.delaware.gov/dhss/dph/files/nitratefaq.pdf.

[205] Schwartz-Nobel, Loretta, Poisoned Nation, p-xix, St. Martins Press, New York, 2007.

[206] Op.Cit. Schwartz-Nobel, p-67.

[207] Kirby, David, *Evidence of Harm*, St. Martins Griffin, New York, 2005.

[208] "Potential Well Water Contaminants and Their Impacts." *EPA*, Environmental Protection Agency, 23 Feb. 2018, www.epa.gov/privatewells/potential-well-water-contaminants-and-their-impacts.

[209] Radio News Report, National Public Radio, March 2011.

[210] Op.Cit. Batmanghelidj, p-122.

[211] Rapp, Doris, *Our Toxic World,* Environmental Medical Research Foundation, Buffalo, New York, 2003.

[212] https://www.vox.com/science-and-health/2017/4/27/15424050/us-underreports-lead-poisoning-cases-map-community.

[213] Op.Cit. Schwartz-Nobel, p-30.

[214] Ibid.

[215] https://www.scientificamerican.com/article/perchlorate-in-drinking-water.

[216] Ibid.

[217] https://www.cancer.gov/about-cancer/understainding/statistics.

[218] Ibid.

[219] Ibid. Environmental Working Group.

[220] Ibid.

[221] Op.Cit. Schwartz-Nobel, p-xvii.

[222] Bragg, Paul C., *The Miracle of Fasting,* Health Science, Santa Barbara, 2010.

[223] Thomas, Patricia, *What's In This Stuff?,* Penguin Company, New York, 2008.

[224] Op.Cit. Robbins, p-xv.

[225] Op.Cit. Schwartz-Nobel, p-65-66.

[226] *An Inconvenient Truth,* Gore, Al, 2007.

[227] Ibid.

[220] Ibid.

CPSIA information can be obtained
at www.ICGtesting.com
Printed in the USA
BVHW042312050723
666779BV00006B/86